DEPARTMENT OF THE ARMY
U. S. Army Corps of Engineers
Washington, D.C. 20314-1000

EM 200-1-4

CEMP-RT

Manual
No. 200-1-4

31 January 1999

Environmental Quality
RISK ASSESSMENT HANDBOOK
VOLUME I: HUMAN HEALTH EVALUATION

1. Purpose. The overall objective of this manual is to provide risk assessors with the recommended basic/minimum requirements for developing scopes of work, evaluating Architect-Engineer (A-E) prepared human health risk assessments, and documenting risk management options associated with Hazardous, Toxic, and Radioactive Waste (HTRW) investigations, studies, and designs consistent with principles of good science in defining the quality of risk assessments. This EM is intended for use by U.S. Army Corps of Engineers (USACE) Project Managers, technical personnel, and contractor personnel.

2. Applicability. This EM applies to all HQUSACE elements and USACE commands responsible for HTRW projects.

3. References. References are listed in Appendix A.

4. Distribution. Approved for public release, distribution is unlimited.

5. Discussion. This manual is intended to provide USACE risk assessors and contractor personnel with supplemental guidance for performance and evaluation of risk assessments under the Comprehensive Environmental Response, Compensation, and Liability Act (CERCLA) as amended by the Superfund Amendments and Reauthorization Act (SARA) of 1986, and the Resource Conservation and Recovery Act (RCRA) as amended by the Hazardous and Solid Waste Amendments (HSWA) of 1984. It is not intended to replace the accepted guidance by the USEPA (e.g., *Risk Assessment Guidance for Superfund, Human Health Evaluation Manual*), but should be used in conjunction with that document. Additional information provided by this manual concerns presentation of the risk assessment results for use in risk management and decision-making, concerns focusing on the decisions, and criteria needed for decisions. Both risk and nonrisk factors are presented for consideration by the risk managers.

FOR THE COMMANDER:

2 Appendices
(See Table of Contents)

ALBERT J. GENETTI, JR.
Major General, USA
Chief of Staff

This manual supersedes EM 200-1-4, Volume I, dated 30 June 1995.

DEPARTMENT OF THE ARMY
U.S. Army Corps of Engineers
Washington, D.C. 20314-1000

EM 200-1-4

CEMP-RT

Manual
No. 200-1-4

31 January 1999

Environmental Quality
RISK ASSESSMENT HANDBOOK, VOLUME I: HUMAN HEALTH EVALUATION

Table of Contents

Chapter   Page   Chapter   Page

CHAPTER 1

1.0 INTRODUCTION

1.1 PURPOSE AND SCOPE

This handbook, *Risk Assessment Handbook: Volume I - Human Health Evaluation*, provides technical guidance to the U.S. Army Corps of Engineers (USACE) risk assessors and risk assessment support personnel for planning, evaluating, and conducting Human Health Risk Assessments (HHRAs) in a phased hazardous, toxic, and radioactive waste (HTRW) response action. The handbook, a compendium to the *Risk Assessment Handbook: Volume II - Environmental Evaluation* (Engineer Manual (EM) 200-1-4), encourages the use of "good science" within the framework of existing U.S. Environmental Protection Agency (USEPA) risk assessment guidelines.

Reference and overview resources:
- Required and Related References (Appendix A)
- Abbreviations and Acronyms (Appendix B)

Risk characterization is a similar process for both human health and Ecological Risk Assessments (ERAs). The fundamental paradigm for human health risk characterization has four phases: (1) hazard identification, (2) dose-response assessment, (3) exposure assessment, and (4) risk characterization. Similarly, the fundamental framework for ecological risk characterization includes four phases: (1) problem formulation, (2) ecological effects characterization, (3) exposure characterization, and (4) risk characterization.

This handbook encourages the concurrent assessment of human and ecological risks so that data collection activities are coordinated and risk managers are provided risk characterization results in a timely manner. Risk characterization results for human and ecological receptors should be reasonable and communicated to the risk managers in a clear and unbiased manner to facilitate the making of balanced and informed risk management decisions.

1.1.1 Objectives. The overall objective of this handbook is to allow the users to be familiar with the risk assessment process so that quality data will be collected and used in preparing a site-specific risk assessment. Specifically, the objectives are:

- To provide guidance for all risk assessments completed under contract with USACE or those for which USACE provides technical oversight (including active Installation Restoration Program [IRP] and Formerly Used Defense Sites [FUDS] and other Federal agencies/facility sites), in compliance with Federal environmental laws and regulations.

- To allow users to be familiar with the application of the data quality design process with respect to conducting risk assessments, so that data collected will support risk assessment conclusions.

- To highlight those decision criteria specific to each phase of HTRW project execution that support risk management decisions.

- To provide minimum requirements for evaluating contractor-prepared risk assessments, assuring that the assessment will adequately support site decisions of an HTRW response action.

- To acknowledge areas of uncertainties where "good science," based on professional judgement and sound scientific principles, is used to determine the need for removal actions or interim measures, further investigation, further action, or no further action (NFA) needed (site closeout).

- To refine understanding of EPA's concepts and application of risk assessment guidelines for site assessment and remediation, especially to support the USACE HTRW program goals.

1.1.2 Scope. This guidance document is not intended to be a "how to" manual which prescribes step-by-step procedures or instructions for preparing an HHRA. Rather, it presents recommendations for scoping, managing, evaluating, and communicating to risk managers and other stakeholders the potential risks posed by hazardous Chemicals Of Concern (COCs) at Comprehensive Environmental Response, Compensation, and Liability Act (CERCLA) sites, Resource

Conservation and Recovery Act (RCRA) sites, and other sites managed under the HTRW program. This handbook provides concepts for performing a risk assessment consistent with "good science" and accepted regulatory procedures. The following areas are not covered in this handbook:

- Biological hazards microbes (natural or genetically engineered) and other biological agents.

- Radioactive hazards - radioactive wastes, radiation generating devices, and radioactively contaminated materials.

- Lead-based paint and asbestos hazards.

- Physical hazards - building demolition/debris removal.

- Study elements and regulatory requirements of a Natural Resource Damage Assessment.

1.1.3   Intended Audience and Use. This document is prepared primarily for use by USACE personnel who are responsible for scoping, directing, and reviewing HHRAs performed for HTRW response action sites. The guidelines provided by this document are consistent with and should be considered in addition to existing EPA guidance contained in the *Risk Assessment Guidance for Superfund (RAGS), Volume I, Part A* (USEPA, 1989j), *Part B* (USEPA, 1991d), *Part C* (USEPA, 1991e), and *Part D* (USEPA, 1998a), and *Data Usability for Risk Assessments* (USEPA, 1992h). The EM entitled *Technical Project Planning (TPP) Process (EM 200-1-2)* (USACE) should be reviewed, particularly for understanding the process described in Chapter 2 of this handbook on how to determine data quality objectives (DQOs) to support a risk assessment.

The data collection, assessment, characterization of risk and uncertainty, and the risk management decision-making (RMDM) aspects presented in this handbook are intended to satisfy RCRA and CERCLA regulatory requirements. The assessment of human health risks under these two functionally equivalent programs is essentially the same. If both regulatory programs are applicable at a site or unit, the risk assessment components should be closely coordinated to avoid duplication of effort. Where possible, the technical

and risk management approaches should be incorporated as specific language in agreements with EPA or states.

1.1.4   Contents of the Handbook. Chapter 1 presents the purpose, scope, concept, science/policy considerations, and the use of risk assessment in HTRW programs. It provides a description of the USACE HTRW program, quality required for performing a risk assessment, and an understanding of how risk assessments serve management decision needs. Relevant Federal statutes/regulations, agency guidance and directives, and state requirements are highlighted in this chapter.

Chapter 2 presents the major scoping or project planning elements under CERCLA as amended by the Superfund Amendments and Reauthorization Act (SARA) of 1986, and RCRA as amended by the Hazardous and Solid Waste Amendments (HSWA) of 1984. Particular emphasis is placed on the early development of a conceptual site model (CSM) in the data quality design process to identify data needs, optimize data collection efforts, and recommend options for site decisions.

Chapter 3 provides an introduction to the HHRA process as it applies to screening-level assessments. Screening-level HHRAs are typically utilized in the Preliminary Assessment/Site Inspection (PA/SI) or RCRA Facility Assessment (RFA) stage of site investigations.

Chapter 4 is intended to provide the risk assessor with the minimum content expected to be included in a Baseline Risk Assessment (BRA), conducted during the Remedial Investigation (RI) or RCRA Facility Investigation (RFI) phase of investigations. This chapter stresses the importance of properly identifying the Chemicals of Potential Concern (COPCs) and developing a thorough understanding of the dynamics or inter-relationships of multiple pathway exposure models. Appropriate methods for estimating exposure point concentrations are also presented. The importance of objectively and realistically characterizing site hazards or risks is discussed relative to satisfying the regulatory requirements of protectiveness of human health and the environment.

Chapter 5 provides the risk assessor with information to evaluate risk assessments conducted during the Feasibility Study (FS) or Corrective Measures Study

(CMS) and Remedial Design/Remedial Action (RD/RA) or Corrective Measures Implementation (CMI) phases of investigations.

Chapter 6 provides guidance on the risk and uncertainty aspects of RMDM. Both risk and non-risk information are collected and presented for consideration by the manager. This chapter emphasizes balancing the need for protection of human health with other project constraints, including the level of confidence and uncertainty in the risk assessment results. It details approaches for evaluating the need for NFA, removal (or interim corrective measure), and remediation. Additionally, Chapter 6 provides the risk assessment information inputs into the decision criteria and rationale for the selection of remedial alternatives or corrective measures. Chapter 6 concludes that the risk assessor is responsible for presenting key risk information to be used as input into risk management options including documentation of uncertainty and rationale.

1.2 USACE ROLE IN THE HTRW PROGRAM

In the execution of USACE environmental missions, the HTRW program is organized and staffed to respond to assignments for the following national environmental cleanup programs:

- EPA Superfund Program (CERCLA)

- Defense Environmental Restoration Program (DERP):

    - IRP
    - FUDS
    - Department of Defense and State Memorandum of Agreement/Cooperative Agreement Program (DSMOA/CA)

- Base Realignment and Closure (BRAC)

- Environmental Compliance Assessment System (ECAS) (USACE 1992a)

- HTRW environmental restoration support for Civil Works projects and other Federal agencies (Department of Defense [DOD] and non-DOD)

For the purpose and intended use of this risk assessment handbook, the focus is on the DERP and BRAC cleanup programs to address CERCLA- and RCRA-related issues.

1.2.1   DERP. DERP, codified in 10 USC Chapter 160, provides central program management for the cleanup of DOD hazardous waste sites consistent with the provisions of CERCLA. The goals of the program are: (1) the identification, investigation, research, and cleanup of contamination from hazardous substances; (2) correction of other environmental damage which creates an imminent and substantial endangerment to the public health or the environment; and (3) demolition and removal of unsafe buildings and structures.

1.2.2   BRAC. BRAC is an environmental restoration program with the mission to restore or clean up DOD installations in preparation of real property disposal or transfer. The Base Closure Account (BCA) funds the BRAC program. The BCA is authorized under the Defense Authorization Amendments and Base Closure and Realignment Act of 1988 and the Defense Base Closure and Realignment Act of 1990. These funds are used to define the nature and scope of contamination, perform RA, and document the condition of real property by issuance of the Finding of Suitability to Lease (FOSL) (DOD, 1993) and the Finding of Suitability to Transfer (FOST) (DOD, 1994a). The Community Environmental Response Facilitation Act (CERFA) (Public Law 102-426) amends CERCLA Section 120(h) and requires Federal agencies to define "real property" on which no hazardous substances and no petroleum products or their derivatives were stored for 1 year or more, were known to have been released, or were disposed of before the property can be transferred. Transfer of contaminated property is allowed as long as the RA to clean up the site is demonstrated to be effective to EPA.

1.2.3   Others. Other components of the USACE HTRW program include:

- EPA Superfund program support - Through an interagency agreement (IAG) and upon EPA request, USACE acts as the Federal government's contracting officer in conducting "Federal Lead" RD and construction activities. USACE may also provide other technical assistance to EPA in support of response actions.

- DSMOA/CA - DOD reimburses states and territories up to one percent of the costs for technical services for environmental restoration cleanups. USACE is responsible for execution of activities which include establishing, managing, implementing, and monitoring the DSMOA/CA program.

- Non-mission HTRW work for others - Through IAGs, non-DOD Federal agencies utilize the technical expertise and experience in work relating to the RCRA, CERCLA, and underground storage tank (UST) investigation and response actions under the HTRW program for non-DOD Federal agencies.

- Guidance for Civil Works projects - The Civil Works districts may request technical support and guidance from HTRW program elements.

1.2.4 HTRW Program Organization. Army Regulation (AR) 200-1 (USA) and USACE HTRW Management Plan (USACE, 1996a) describe the USACE organizational elements in support of DERP, BRAC, and other programs. Their major responsibilities include, but are not limited to, the following:

- The Assistant Secretary of the Army for Installations, Logistics, and the Environment (ASA [I,L,E]).

- Headquarters, U.S. Army Corps of Engineers (HQUSACE) - The Military Programs Directorate - Environmental Restoration Division (CEMP-R) develops, monitors, coordinates, and proposes program management policies and guidance, and provides funding and manpower requirements to the program customers.

- The Director of Environmental Programs (DEP) within the office of the Assistant Chief of Staff for Installation Management (ACSIM) is responsible for interfacing with Department of the Army (DA) components for policies and funds for IRP/FUDS/BRAC executed by USACE.

- HTRW Center of Expertise (CX) is primarily responsible for maintaining state-of-the-art capability, providing technical assistance to other USACE elements, providing mandatory review of designated HTRW documents, and as requested, providing technical and management support to HQUSACE.

- Ordnance and Explosives (OE) CX is primarily responsible for maintaining state-of-the-art technical capabilities in OE, performing SIs, Engineering Evaluations and Cost Analyses (EE/CAs), and removal design phases of OE projects.

- Divisions are responsible for providing program oversight of all HTRW environmental restoration projects and designating project management assignments for HTRW projects.

- HTRW design districts provide the Division Commander with technical support in the areas of health and safety, chemical and geotechnical data quality management, environmental laws and regulations, risk assessment, contracting and procurement, and technical design and construction oversight.

- Geographic districts are responsible for managing the execution of RAs as well as PAs, removal design, and removal action related to the FUDS program.

1.3 OVERVIEW OF HTRW RESPONSE PROCESS

HTRW response actions involve all phases of a site investigation, design, remediation, and site closeout. The HTRW response action process is phased and performed in accordance with EPA procedures for assessing uncontrolled hazardous waste sites under CERCLA or RCRA. The following sections generally describe the CERCLA and RCRA processes, which are functionally equivalent to one another in objectives and types of site decisions to be made throughout each process.

1.3.1 CERCLA Process. CERCLA, commonly known as "Superfund," establishes a national program for responding to uncontrolled releases of hazardous substances into the environment. The regulation implementing CERCLA is the *National Oil and Hazardous Substances Pollution Contingency Plan* (NCP) (USEPA, 1990c). In general, the CERCLA process consists of the site assessment phase and the remedial phase as described below; however, removal actions (as allowed by the NCP) may be taken at any time during the CERCLA process. It should be noted that the general framework established under the CERCLA process has been adopted for use in environmental cleanup under other programs, e.g., the cleanup of petroleum, oil, and lubricants (POLs)[1] at FUDS or active installations not listed on the proposed or final National Priorities List (NPL). Therefore, certain CERCLA project phases described below (specifically, the Hazard Ranking System [HRS], NPL, and site deletion), are not applicable to these types of sites.

1.3.1.1 Site Assessment Phase - To Identify Sites for Further Evaluation.

- **Site Discovery** - EPA identifies and lists in the CERCLA Information System (CERCLIS) possible hazardous substance releases to be evaluated under Superfund.

- **PA** - While limited in scope, a PA is performed on sites listed in CERCLIS to distinguish sites which pose little or no threat to humans and the environment and sites that require further investigation or emergency response.

- **SI** - An SI identifies sites which (1) have a high probability of qualifying for the NPL or pose an immediate health or environmental threat that requires a response action, (2) require further investigation to determine the degree of response action required, and/or (3) may be eliminated from further concern.

- **HRS** - At the end of both the PA and SI, EPA applies a scoring system known as the HRS to determine if a site should receive a "no further remedial action planned" recommendation or be listed on the NPL for further action. An HRS can also be used to support other site evaluation activities under CERCLA (see *The Revised Hazard Ranking System*, USEPA, 1992a). Although HRS scoring is the EPA's responsibility, site investigations should be designed

in such a way as to assure that adequate data is available for EPA to perform the scoring.

- DOD has developed the *Relative Risk Site Evaluation Primer* (1994b) to rank sites primarily for resource allocation and program management purposes. Although neither a replacement nor alternative for HRS scoring, this model suggests that stakeholders consider evaluation factors (contaminant hazard factor, migration pathway factor, and receptor factor) to categorize sites according to "high," "medium," and "low."[2]

- **NPL** - Sites placed on the NPL (based on an HRS score of 28.5 or greater, state nomination, issuance of a health advisory by the Agency for Toxic Substances and Disease Registry (ATSDR), or other method) are published in the Federal Register and are eligible for Superfund-financed RA. DOD sites on the NPL, although not eligible for Superfund-financed RA, are eligible for Defense Environmental Restoration Account (DERA)-funded response actions.

1.3.1.2 Remedial Phase - To Determine the Degree of Risk Based on Nature and Extent of Contamination and Implement Cleanup Remedies if Warranted.

- **RI** - The RI is a field investigation to characterize the nature and extent of contamination at a site and implement cleanup remedies if warranted. A BRA, which includes both a HHRA and an ERA, is performed as part of the RI. The BRA is a component of the RI/FS report.

- **FS** - Based on data collected during the RI[3], remedial alternatives are developed, screened, and analyzed in detail. After potential alternatives are developed, they are screened against three broad

---

[1] POLs are not listed as hazardous substances under CERCLA and therefore are not subject to CERCLA response actions. However, unless the state has specific requirements for remediating POL sites, the CERCLA process may be utilized to address the site.

[2] The *Relative Risk Site Evaluation Primer* (DOD 1994b) has replaced the Defense Prioritization Model, which has features comparable to the HRS.

[3] If the BRA contained in the RI indicates that risks are acceptable or insignificant, the FS will not be done and the site will be closed out.

criteria: effectiveness, implementability, and cost. Those alternatives which pass this initial screen will be further evaluated according to EPA's nine criteria[4] and other risk management considerations not included in the criteria (e.g., environmental justice under Executive Order (EO) 12898) before one or more of such remedies is proposed for selection.[5]

- **Proposed Plan/Record of Decision (ROD)** - After the RI/FS process has been completed, a Proposed Plan is made available for public comment. The Proposed Plan identifies the remedies for the site jointly selected by the lead agency and the support agencies, and indicates the rationale for the selection. All final decisions and response to public comments are entered in a legal administrative record, the ROD.

- **RD/RA** - RD is a subactivity in remedial implementation where the selected remedy is clearly defined and/or specified in accordance with engineering criteria in a bid package, enabling implementation of the remedy. RA is a subactivity in remedial response involving actual implementation of the selected remedy.

- **Five Year Review/Site Deletion** - Upon completion of all RAs, CERCLA and the NCP allow for the reclassification or deletion of the site from the NPL. If an RA results in any hazardous substances remaining on site, CERCLA Section 121(c) requires a review of the remedy once every 5 years to assure that: (1) the site is maintained, i.e., the remedy (including any engineering or institutional controls) remains operational and functional; and (2) human

---

[4] The nine criteria are: (1) overall protection of human health and the environment; compliance with Applicable or Relevant and Appropriate Requirements (ARARs); (3) long-term effectiveness/permanence; (4) short-term effectiveness; (5) reduction of toxicity, mobility, or volume; (6) implementability; (7) cost; (8) state acceptance; and (9) community acceptance.

[5] If the RI shows no unacceptable risk, regulators may agree to eliminate the FS and proceed directly to a no-action proposed plan.

health and the environment are protected, i.e., the cleanup standards (based on risk or ARARs) are still protective.

1.3.1.3 Removal Action - To Prevent, Minimize, Stabilize, or Mitigate Threat to Humans and the Environment.

CERCLA Section 104 Removal Actions can take place at anytime during the entire CERCLA process. Unlike RAs, removal actions are not designed to comprehensively address all threats at the site. Removal actions may be emergencies (within hours of site discovery), time-critical (initiated within 6 months), non-time-critical (planning for the removal action takes 6 months or longer), or early actions. EE/CAs, comparable to FSs, are required for removal actions that are deemed non-time-critical.

1.3.2 RCRA Corrective Action Process. RCRA requires corrective action for releases of hazardous waste or hazardous waste constituents from Solid Waste Management Units (SWMUs) at hazardous waste Treatment, Storage and Disposal (TSD) Facilities with a permit and those seeking a RCRA permit or approval of final closure. The owner or operator of a facility seeking a RCRA permit must:

- Institute corrective action as necessary to protect human health and the environment from all releases of hazardous waste, and hazardous constituents from any SWMU at the facility.

- Comply with schedules of compliance for such corrective action.

- Implement corrective actions beyond the facility boundary.

The corrective action process has four main components: an RFA, an RFI, a CMS, and a CMI.

- **RFA** - An RFA is designed to identify SWMUs which are, or are suspected to be, the source of a release to the environment. The RFA begins with a preliminary review of existing information on the facility, which may be followed by a visual site inspection. The RFA will result in one or more of these actions: (1) NFA is required, (2) an RFI is to be conducted to further investigate the documented or suspected releases, (3) interim measures are necessary to protect human health or the environment, and (4) referral to other authorities to address problems related to permitted releases.

- **RFI** - An RFI may be required based on the outcome of the RFA. An RFI is accomplished through either a permit schedule of compliance or an enforcement order. The extent of the investigation can vary widely since the investigation site may encompass a specific SWMU or a larger area of concern (AOC) that includes several SWMUs. The RFI results will effect one or more of these actions: (1) NFA is required, (2) CMS is necessary, (3) interim corrective measures are necessary, or (4) referral to another authority to address problems related to permitted releases.

- **CMS** - A CMS is an "engineering evaluation" designed to evaluate and recommend the optimal corrective measure(s) at each SWMU where contaminant levels exhibit unacceptable risks. Medium-specific cleanup levels protective of human health and ecological receptors are developed, and the boundaries or point(s) of compliance are set. At this project phase or before the CMI phase, RCRA provides the designation of an AOC in which remediation wastes may be moved and managed (according to the approved corrective measures) without triggering land disposal restriction regulations under 40 CFR Part 268. Note that a typical CMS is more focused than is usually done for CERCLA FSs. The remedy selected from all potential remedial alternatives, including the "NFA" alternative, should be based on four criteria:

  - Protection of human health and the environment.

  - Attainment of media cleanup standards.

  - Control of sources to eliminate harmful releases.

  - Compliance with RCRA's waste management and disposal requirements.

- **CMI** - A CMI includes the actual design, construction, operation, maintenance, and periodic evaluation of the selected corrective measures.

EPA can impose interim corrective measures on RCRA facilities under corrective action to protect human health and the environment. The interim corrective measures can be taken at any time during the corrective action process.

EPA is accelerating cleanups at RCRA corrective action sites by promoting the reduction of exposure and further releases of hazardous constituents until long-term remedies can be selected. These accelerated cleanup actions are known as "Stabilization Initiatives" (USEPA, 1992n) and are similar in concept and application to the Superfund Accelerated Cleanup Model (SACM) under CERCLA (USEPA, 1992g).

1.3.3   Functional Equivalency of the CERCLA and RCRA Processes. The RCRA and CERCLA programs use different terminology, but follow parallel procedures in responding to releases. In both programs, the first step after discovery of a site is an examination of available data to identify releases needing further investigation. This step is called PA/SI in the CERCLA process and RFA in the RCRA process. If imminent human health and/or environmental threats exist, a mitigating action is authorized, known as a removal action under CERCLA Section 106 or an interim measure under RCRA Section 7003 or 3005(c)(3). Both programs require an in-depth characterization of the nature, extent, and rate of contaminant releases, called an RI in the CERCLA process and an RFI in the RCRA process. This is followed by a formal evaluation and selection of potential remedies in the FS (CERCLA) or CMS (RCRA) project phase. The selected remedy is executed by a RD/RA under the CERCLA process or CMI under the RCRA process. A specific discussion of the functional equivalency of both programs is presented in the preamble discussion of the July 27, 1990 proposed rules for Corrective Action for SWMUs at Hazardous Waste Management Facilities. A diagram comparing the RCRA and CERCLA processes is presented in Figure 1-1.

1.3.4   Role of Risk Assessment in the HTRW Process. Risk assessment has been consistently used as a decision-making tool in one or more steps in the

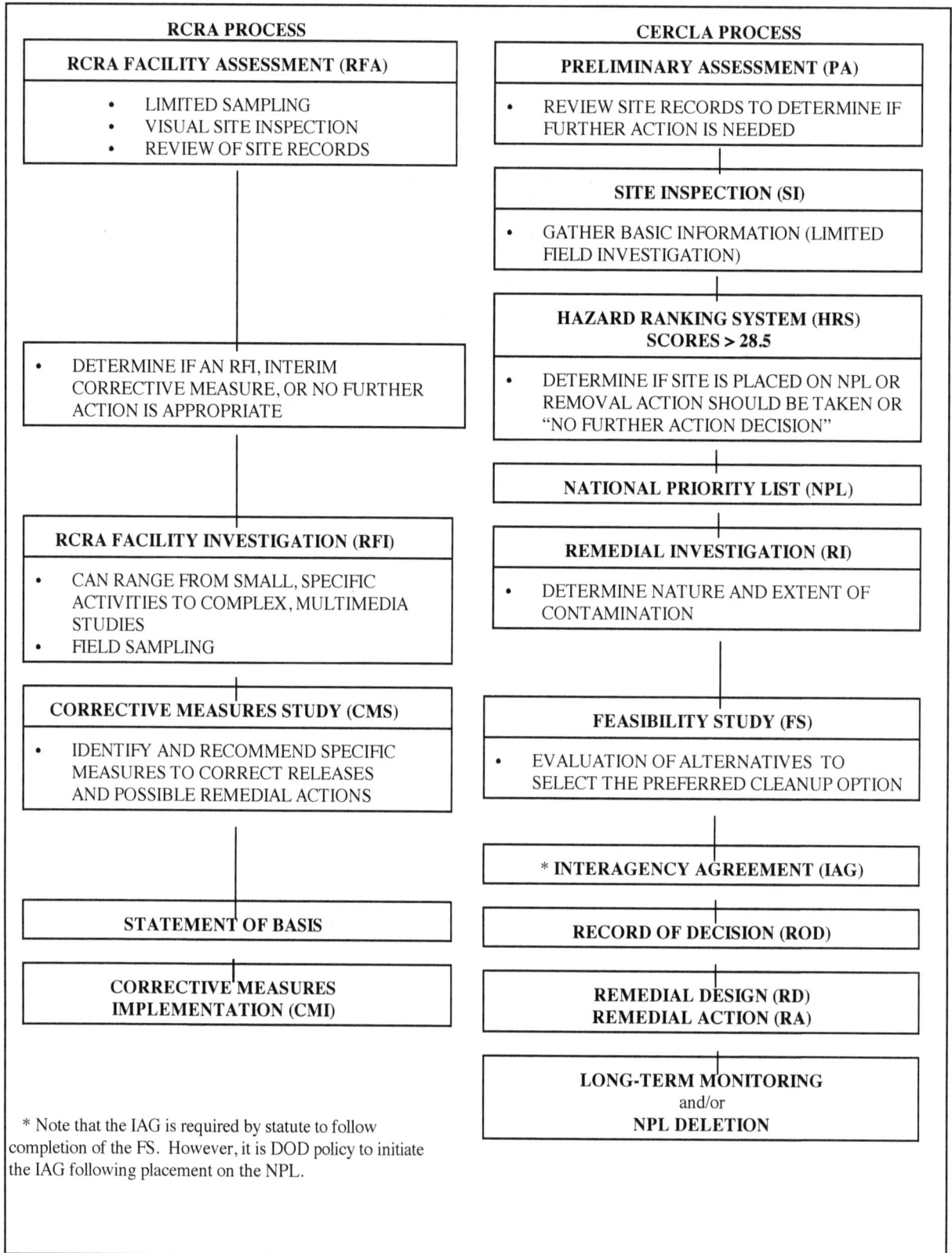

RCRA PROCESS	CERCLA PROCESS
RCRA FACILITY ASSESSMENT (RFA) • LIMITED SAMPLING • VISUAL SITE INSPECTION • REVIEW OF SITE RECORDS	**PRELIMINARY ASSESSMENT (PA)** • REVIEW SITE RECORDS TO DETERMINE IF FURTHER ACTION IS NEEDED
	SITE INSPECTION (SI) • GATHER BASIC INFORMATION (LIMITED FIELD INVESTIGATION)
• DETERMINE IF AN RFI, INTERIM CORRECTIVE MEASURE, OR NO FURTHER ACTION IS APPROPRIATE	**HAZARD RANKING SYSTEM (HRS) SCORES > 28.5** • DETERMINE IF SITE IS PLACED ON NPL OR REMOVAL ACTION SHOULD BE TAKEN OR "NO FURTHER ACTION DECISION"
	NATIONAL PRIORITY LIST (NPL)
RCRA FACILITY INVESTIGATION (RFI) • CAN RANGE FROM SMALL, SPECIFIC ACTIVITIES TO COMPLEX, MULTIMEDIA STUDIES • FIELD SAMPLING	**REMEDIAL INVESTIGATION (RI)** • DETERMINE NATURE AND EXTENT OF CONTAMINATION
CORRECTIVE MEASURES STUDY (CMS) • IDENTIFY AND RECOMMEND SPECIFIC MEASURES TO CORRECT RELEASES AND POSSIBLE REMEDIAL ACTIONS	**FEASIBILITY STUDY (FS)** • EVALUATION OF ALTERNATIVES TO SELECT THE PREFERRED CLEANUP OPTION
	* INTERAGENCY AGREEMENT (IAG)
STATEMENT OF BASIS	**RECORD OF DECISION (ROD)**
CORRECTIVE MEASURES IMPLEMENTATION (CMI)	**REMEDIAL DESIGN (RD) REMEDIAL ACTION (RA)**
	LONG-TERM MONITORING and/or **NPL DELETION**

* Note that the IAG is required by statute to follow completion of the FS. However, it is DOD policy to initiate the IAG following placement on the NPL.

　　　Figure 1-1. Comparison of RCRA and CERCLA processes.

CERCLA and RCRA corrective action processes. A risk screening analysis is used during the PA/SI to determine whether a site may be eliminated from further concern or requires further study, which may be focused on specific areas of the site. A BRA is conducted in the RI. Section 105 of CERCLA/SARA charges the On-Scene Coordinator or Remedial Project Manager (RPM) with the responsibilities of identifying potential impacts on public health, welfare, and the environment, and setting priorities for this protection which is delegated to DOD under Section 115 and EO 12580 for DOD facilities. RCRA Section 3019 requires the facility owner/operator to submit an Exposure Information Report (EIR) which provides exposure and health assessment information for certain storage and land disposal waste management units. In the RFI, as required by permit conditions or enforcement actions under RCRA Sections 3008(h), 7003, and/or 3013, a Health and Environmental Assessment (HEA) is used to determine quantitatively if the site or any of its units has exceeded established health criteria. As indicated in the RFI guidance (USEPA, 1989f), a site-specific risk assessment will be performed prior to the CMS to assess potential risk to humans and to determine if no response action is appropriate. Under CERCLA Section 120, the BRA is one of the primary documents identified for submission to EPA for comment and review in the Federal Facility Agreement (FFA).

Risk assessment in reverse is used to develop risk-based Remediation Goals (RGs) under CERCLA or Target Cleanup Levels (TCLs) (CERCLA Section 121) or Alternate Concentration Limits (ACLs)[6] under RCRA (40 CFR 264.94 and 264.100). Risk-based RGs, TCLs, or ACLs should be developed after the BRA has been performed incorporating site-specific factors in the calculations. Preliminary Remediation Goals (PRGs), corrective action levels, or soil screening levels can be developed at any time in the site investigation process, to determine whether further action is appropriate and to help focus subsequent studies on significant pathways of exposure. The summary or conclusions of the RI BRA,

development of RGs based on allowable exposure, and analysis of alternatives (based on risk and the other criteria) are part of the FS report (USEPA, 1988i).

To be protective of human health, interim corrective measures or remedial alternatives must also be evaluated based on their ability to reduce site risk and their potential impact to humans during and after remediation. This risk evaluation of remedial alternatives is part of the remedy or corrective measure selection process prior to RD/RA (CERCLA Section 121, NCP Section 300.430(e)(1)), and Proposed RCRA Corrective Action Rule, Section 264.525(b)(55 FR 30798, July 27, 1990 and 61 FR 19431, May 1, 1996).

Performing a risk assessment is an iterative process. Risk assessment information is continuously being collected during the HTRW site investigation process, leading to the characterization of risks and uncertainties qualitatively or quantitatively. Risk assessment information is used during various stages of the HTRW site decision process as described below:

1.3.4.1 PA/SI, RFA, or Other Preliminary Site Investigation Activities. In this phase of the site investigation process, risk assessment information is used to determine whether a site may be eliminated from further concern, to identify emergency situations which may require immediate response actions/interim corrective measures, to assess whether further site investigations are required, to develop a data collection strategy, and to set site priority (e.g., to rank sites).

It is important that the limited information gathered in this phase support the risk screening analysis and the HRS scoring if further site investigations are required. Accurate site information should be made available to the ATSDR in an attempt to avoid having health consultations or an advisory issued for the site by ATSDR based on inaccurate site information.[7]

---

[6] ACLs are allowable for ground water contamination only and do not address contamination of other media. Cleanup levels for surface water, sediment and soil are determined utilizing risk assessment as is done in CERCLA.

[7] Under CERCLA Section 104(j)(6), ATSDR is required to conduct health assessment under this Section for sites where individuals may have been exposed to a hazardous substance for which the source is related to a CERCLA release. Health assessments are generally based on SI, RI, Superfund risk assessment (human health evaluation), and studies submitted to ATSDR. In addition, ATSDR may conduct an analytical investigation that evaluates the

1.3.4.2 RI, RFI, or Other Additional Site Investigation Activities. In this phase of the site investigation process, existing chemical data and other exposure information are generally available. Data collected in this phase should comprise those media and pathways identified in the preliminary screening, including background data. If the data are useable and appropriate for the potential exposure pathways considered to be complete, baseline risks can be estimated. The results of the risk assessment will be used in the FS to determine the degree of response action required. RAs should be initiated to address the risks associated with an operable unit (OU), a SWMU, an area of contamination (AOC), an area of interest/concern, or an exposure area or unit.

An OU, as defined in the NCP, "is a discrete action that comprises an incremental step toward comprehensively addressing site problems." OUs provide a procedural basis for phasing multiple control measures that make up an RA, which may be used as a construction management tool during installation of complex RA, and which can provide manageable geographic areas for study. Areas of a site which are concerned with a specific receptor group may be used as the basis of OU designation which allows for effective evaluation of exposure pathways, simplifying the risk assessment of the site into manageable components. DOD facilities are much larger than traditional Superfund sites, and designation of OUs is an important part of designing the risk assessment to effectively define RA requirements.

To avoid triggering RCRA land disposal restrictions or minimum technology requirements, OUs may be combined to form an AOC for the purpose of implementing response action. A similar concept has been applied for combining SWMUs. It should be noted that the BRA completed in the RI serves to identify the need for response action and the relative degree of response required based on protection of human health and the environment.

1.3.4.3 FS, RD/RA, CMS/CMI, or Other RD and Implementation Activities. During the feasibility, treatability, or other remedial measure study phase, an evaluation of short-term and long-term risks associated with remedial alternatives is required under CERCLA Section 121, as is the development of cleanup levels.

Risk-based RGs/TCLs/ACLs can be derived based on EPA-established procedures (e.g., RAGS Parts A and B).[8] Specifically, risk assessment will be used to select a remedy by comparing among the alternatives the potential human/environmental impact during remediation (short-term and long-term) and the residual risks after remediation. This comparative analysis can be performed qualitatively for the ability of the alternatives to achieve the RGs, TCLs, ACLs (along with other criteria such as cost and long-term effectiveness). A more effective approach for many sites will be to perform quantitative evaluation of the risks associated with each remedial alternative or corrective measure, based on the alternative's long-term and short-term impact on risk to receptors. All potential receptors during and after the RA periods should be considered.

As with environmental monitoring, risk assessment can play a key role in assessing the residual risks and to establish ACLs. It can be used as a measuring tool to gauge the success of the RAs or corrective measures. See RAGS, Part C (USEPA, 1991e), *Alternate Concentration Limit Guidance Part 1 - ACL Policy and Information Requirements* (EPA 1987b), and *Alternate Concentration Limit Guidance Based on 264.94(b) Criteria, Case Studies* (EPA 1988f).

1.3.4.4 Use of Risk Assessment in Special Studies. Risk assessment techniques are used in virtually all phases of CERCLA, RCRA, and other HTRW processes. Therefore, risk assessment should be planned for and conducted to provide input to discussions associated with each phase. There are also special studies in addition to

---

possible causal relationships between exposure to hazardous substances and disease outcome by testing a scientific hypothesis. Exchanges of information and reports with ATSDR will be coordinated through the U.S. Army Center for Health Promotion and Preventive Medicine (USACHPPM).

---

[8] This manual emphasizes the need for careful HTRW project planning for adequate data collection to support a site decision. Risk assessment is a powerful decision tool; yet, misapplication of risk assessment procedures and concepts and poor data quality and quantity could lead to inaccurate assessment of risk and may lead to incorrect or poor site decisions.

executable phases discussed previously, specifically, protectiveness or "How clean is clean?" The following are examples of risk assessment in special studies:

- ARAR waiver - EPA has indicated that a non-zero Maximum Contaminant Level Goal (MCLG) is an - ARAR in the remedy selection. MCLG does not take into account specific site conditions and exposure patterns or economic and technical feasibility of implementation. Even though the non-zero MCLG may be considered an ARAR, a risk assessment can be used to evaluate the appropriateness of the non-zero MCLG. If a site-specific alternative cleanup level is as protective, an ARAR waiver request may be submitted under CERCLA Section 121(d)(2). The same process may be used to waive state ARARs, some of which are based on aesthetics including sight and odor.[9]

- Emergency response - The effectiveness of a proposed removal action, particularly for non-time-critical response action, should be evaluated in terms of the ability of the response action to reduce exposure. A screening risk assessment can be conducted to evaluate the response actions for relatively straightforward sites, although a BRA may be more appropriate for complex sites and cost recovery actions. This is particularly critical since EPA and some states want to implement early actions and presumptive remedies for certain sites. USACE HTRW risk assessment staff and design districts should consider all options, based on effectiveness of the action, and other criteria in the risk reduction efforts.

- Compliance with state air programs - CERCLA and RCRA sites are potential sources of air emissions. These air emissions may be present before and/or during the response action (removal or remediation), or during the operation and

---

[9] EPA has compiled thresholds for odor for chemicals based on an extensive literature search. The updated odor thresholds should be consulted to evaluate if the ARAR (if based on odor) is reasonable. See *Reference Guide to Odor Thresholds for Hazardous Air Pollutants Listed in the Clean Air Act Amendments of 1990*, (USEPA, 1992f).

maintenance of the response action. Of particular concern are volatile and semivolatile organic chemicals, particulate matter, heavy metals, and acids. Operations implemented during the cleanup process (i.e., RI, removal action, or construction of a selected remedy) may emit air pollutants. Examples of operations which may act as a source of air emissions include soil handling, air stripping, onsite incineration, and equipment used in solidification/stabilization processes. USACE risk assessors should consult with state air regulatory personnel to determine the exact risk assessment requirements for evaluating air pathway exposures within that state. If potential risks are determined following state guidelines, resulting requirements for air emission limitations or emission control technologies should be discussed with the appropriate USACE personnel on the RD team.

- Risk assessment will be useful to assess the impact of the response actions (new sources) and the baseline condition (an existing source), for attainment of the National Ambient Air Quality Standards and substantive requirements embodied in the State Implementation Plan (SIP). See *ARAR Fact Sheet - Compliance with the Clean Air Act and Associated Air Quality Requirements* (USEPA, 1992l).

1.4 CONCEPT OF RISK ASSESSMENT AND GOOD SCIENCE

Risk assessment can be qualitative or quantitative. It includes an integration of hazard (dose and response), exposure (intake), and characterization of the potential risks/hazards and uncertainties. The process relies on strong fundamental scientific principles; the management aspect relies on application of policy as well as professional judgment and experience. This view is reflected by the National Academy of Sciences (NAS) and EPA who recognized the inherent uncertainties in the risk assessment methodologies. The uncertainties are primarily caused by various unknowns in the risk estimate calculation, which, in many cases, requires making assumptions relating to predictive modeling or inferences of certain scientific principles (Federal Focus

Inc., 1991).[10] This paragraph highlights the principles, instructions, or recommendations of assessing the impact on human health from chemicals in the environmental media at HTRW sites.

1.4.1 Basic Concepts. The fundamental principles of good science and quality entail the thorough understanding of: (a) site chemical data; (b) an understanding of site-related and background risks; (c) physical, chemical, and toxicity information associated with site chemicals; (d) fate and transport of site chemicals; (e) intake and extent of absorption; (f) the dose-response relationship of site chemicals; (g) uncertainties and limitations of the derived risk estimate;

---

[10] There can be significant uncertainties in the input parameters used in the risk assessment model, assuming that the model is the best scientific representation which can be used to predict potential health consequences from the exposure to chemicals in the environment. Since these models are used to support site decisions and policy-making, quantitative examination of these uncertainties is important. Presentation of risk estimates under the average and reasonable maximum exposures (RME) is now required by EPA's Superfund office. Recently, there has been an increased use of Monte Carlo (MC) analysis to propagate uncertainties through repetitive risk assessment calculations. Two examples of the application of MC are: (1) to determine a more accurate estimate of "reasonable maximum" risk than the use of standard default (normally high end) values for exposure input factors which could magnify risks to the Theoretical Upper Bound Estimate region of the risk probability curve, and (2) to evaluate the trade-off between extent (and thus cost) of remediation and degree of confidence in achieving adequate protection of health. MC can be used to provide risk estimates based on simulations of only a few key parameters which could substantially impact risks. These parameters are normally identified by performing a sensitivity analysis which compares the relative impact on the risk estimates (ranges) associated with each input parameter's maximum and minimum values while keeping other parameter values unchanged. There are off-the-shelf computer software programs for MC analysis in risk assessment, e.g., Crystal Ball®, At-Risk®, and others.

and (h) the best approach to characterize risk objectively.

The application of good science or definition of quality in the risk assessment reduces or defines uncertainties in a risk assessment. This application results in an unbiased risk characterization which allows risk managers to make informed site decisions. If the risk assessment uncertainties are well documented, and the results presented in a manner which can be easily understood by decision-makers, then this element of decision-making has more meaning relative to the other elements of risk management.

1.4.2 Risk Assessment as Decision Criteria in the HTRW Program. The role of a risk assessment in the site decision-making process at CERCLA and RCRA Corrective Action sites has been well defined by EPA either through rule-making or program directive/guidance. Therefore, risk assessments have been used as decision criteria in the USACE's HTRW program involving CERCLA and RCRA sites. For BRAC, FUDS, or other HTRW work which may not be on the NPL, risk assessments should be similarly applied. Activities at these sites require the evaluation of potential health and environmental risks in order to return the property to conditions appropriate for the current and planned future land uses. Therefore, a site-specific BRA is an important decision tool to USACE customers. If cleanup is needed, the extent or level of cleanup required will be based on results of the BRA, in addition to ARARs or other non-risk factors. Therefore, risk assessment is used as a decision tool at all HTRW response action sites.

1.5 POLICY CONSIDERATIONS AND RISK MANAGEMENT

This section presents a general discussion of the influence of policy considerations in risk assessment and risk management. Because of the implications of policy considerations on the site decision process, the risk assessors and risk managers are encouraged to identify the policies early in the decision process.

Unlike regulations which are enforceable, policies or published guidelines are administrative procedures or requirements concerning certain environmental regulations. DOD has issued directives to components (Army, Navy, Air Force, and Defense Logistic Agency), reaffirming DOD's commitment to comply with specific environmental laws or EOs. The respective components have also issued directives or orders expressing the same procedures or requirements. USACE will follow such policies or directives issued by DOD or its components regarding compliance with Federal environmental laws in the execution of HTRW response action at DOD installations or facilities. Some states or regional environmental control boards have also issued environmental policies or guidance. In the unlikely event that a policy is scientifically incongruent with site situations, early identification and resolution are critical. HQUSACE or HTRW CX technical staff should be consulted in these instances. All major policies used in making site decisions should be identified in the ROD or site decision documents so that the USACE customers and other stakeholders can judge the merit of these policies in achieving protection of human health and the environment.

1.5.1 Relationship Between Policy Considerations and Risk. A risk assessment is the technical evaluation of the degree of hazard or risk associated with exposure to contamination of an environmental medium or media. Risk management is oriented toward deciding whether RAs are warranted in light of the results of a risk assessment. The NAS National Research Council (NRC) defines risk management as "the process of weighing policy alternatives and selecting the most appropriate regulatory action, integrating the results of risk assessment with engineering data and with social, economic and political concerns to reach a decision" (NRC, 1983). NAS has identified four key components in managing risk and resources: public participation, risk assessment, risk management, and public policy decision-makers (NRC, 1994).

In making risk management decisions, the risk manager considers the degree of risk, technical feasibility to address risk, costs and benefits, community acceptability, permanence of the proposed actions, and other similar factors which are subject to policy considerations or regulatory requirements. As such, risk management is an important part of the USACE HTRW site response process, as it combines results of the risk assessment, regulatory requirements, and applicable agency policies (e.g., applicable DOD policies for defense sites).

1.5.2 USACE Policy Considerations. In an effort to standardize risk assessment procedures within the USACE HTRW program, the following considerations should be consistently applied to all site-specific risk assessments. Although not designated as DA or USACE policy at this time, these issues are based on sound science and will assist in making risk management decisions. At the appropriate locations within the text (see paragraph references below), these policy considerations are presented in bold typeface within double outlined text boxes, including implementation directives, as required.

• The risk assessment shall be given, at a minimum, equal consideration with other factors in the risk management decision. See Paragraph 6.1.

• All risk assessments shall include a statistically robust, significant, and defensible set of background concentrations. See Paragraph 4.3.3.2.2.

• Future land uses for risk assessment purposes and for development of remedial action objectives (RAOs) shall be land uses that are reasonably expected to occur at the site or facility. See Paragraph 4.4.4.

• If the cumulative site risk calculated in the risk assessment does not exceed 1E-04 for reasonable exposure scenarios, ARARs are not exceeded, and ecological impacts are not significant, no RA should be required. See Paragraph 6.2.2.

• The exposure assessment of a risk assessment shall utilize site-specific frequencies and durations whenever possible. A minimum of two risk estimates should be presented for each land use scenario, the RME and the central tendency (CT). See Paragraphs 4.4.5.1.3 and 4.4.5.1.6.

• Use of the EPA's Integrated Exposure Uptake and Biokinetic (IEUBK) model for lead exposures should be limited to residential, childhood exposures only. Where non-residential exposures are expected, an adult lead intake model should be used. See Paragraph 4.5.7.1.2.

• RGs must be developed and applied in the context of exposure area and exposure point concentrations. It is unnecessary to remediate all media to or below the RG. See Paragraph 5.2

1.5.3 EPA Headquarters, Regional and State Policies. To successfully complete a risk assessment for use in making site decisions, HTRW project managers (PMs) and risk assessors generally work with Federal, regional, and state regulatory agencies to identify their specific policies or procedural requirements. HTRW risk assessors should identify and assist, where appropriate, in negotiations with the agencies on policies, procedures, and assumptions which are questionable.

All HTRW response actions should be in compliance with the Regulatory Policy Guideline issued under EO 12498 (1985), which states, "Regulations that seek to reduce health or safety risks should be based upon scientific risk assessment procedures, and should address risks that are real and significant rather than hypothetical or remote." Whenever possible, USACE's HTRW position should be supported by scientific principles, site data, or literature values. USACE recognizes that at times, agencies have to set policies in the absence of scientific consensus; however, USACE, through the HTRW program, is responsible for applying such policies properly and objectively based on site-specific considerations.

1.5.4 Risk-Based Management Decisions for Site Actions. Risk managers select the most appropriate remedy by considering "trade-offs" among different remedial alternatives and evaluating the ability of the alternatives to accomplish the overall project objectives. To improve the quality of risk-based management site decisions, HTRW risk assessors should identify key information that can affect that decision-making. This information should include policy considerations, assumptions concerning the margins of safety, and the use of other relevant data not associated with the site in the risk assessment. The sources of such policies and data, as well as the qualifications of persons/organizations recommending the policies or use of data, should be clearly identified. HTRW risk assessors can further help risk managers by providing an explanation of uncertainties in the risk assessment. When science deviates from policies or assumptions inherent in the risk assessment, it is the responsibility of

HTRW risk assessors to clearly identify these instances as potential uncertainties as well.

1.6 REGULATORY DIRECTIVES AND GUIDANCE

This section highlights major EOs, Federal statutes/regulations under which the HTRW programs operate, and EPA risk assessment guidelines which provide the basis for development of this handbook. Irrespective of the procedures or mechanics for conducting risk assessments according to regulatory guidelines, all risk assessments performed under the HTRW response action must be based on "good science" and reasonable and unbiased scientific judgment. Although this section lists only major applicable EOs and directives, others may be accessed through the appropriate agencies and databases on the Internet.

1.6.1 EOs and Federal Statutes/Regulations.

EO 12088 (1978), *Federal Compliance with Pollution Control Standards*, established the mechanism by which the Executive Branch assures that its facilities (in various departments) meet their compliance responsibilities by complying with substantive and procedural requirements of Federal environmental statutes. These statutes include: Endangered Species Act (ESA); the Clean Air Act (CAA); the Federal Water Pollution Control Act (Clean Water Act [CWA]); the Solid Waste Disposal Act (as amended by RCRA); the Noise Control Act; the Marine Protection, Research and Sanctuaries Act (Ocean Dumping Act); the Safe Drinking Water Act (SDWA); the Toxic Substances Control Act; the Federal Insecticide, Fungicide, and Rodenticide Act; and the National Historic Preservation Act.

EO 12498 (1985), *Government Management*, incorporates by reference the regulatory principles contained in a Task Force report regarding future significant regulatory actions. Two principles of interest are:

• Regulations that seek to reduce health or safety risks should be based upon scientific risk-assessment procedures, and should address risks that are real and significant, rather than hypothetical or remote; and

- To be useful in determining overall benefits and costs, risk assessments must be scientifically objective and include all relevant information. In particular, risk assessment must be unbiased best estimates, not hypothetical "worst cases" or "best cases." In addition, the distribution of probabilities for various possible results should be presented separately, so as to allow for an explicit "margin of safety" in final decisions.

EO 12580 (1987), *Superfund Implementation*, requires all Federal agencies to comply with CERCLA/SARA and NCP in the same manner as the private sector. This Order delegated to the Secretary of Defense the response authority of DOD, which includes removal/RAs, site investigation and risk assessment, remedy selection, performance of PAs, and assuming natural resource trustee's responsibilities for current and former DOD facilities, and others. The Office of the Deputy Under Secretary of Defense for Environment Security (ODUSD[ES]) is responsible for carrying out the Secretary's responsibilities and administering the DERP in compliance with this Order.

EO 12777 (1991), *Implementation of Section 311 of the Federal Water Pollution Control Act of October 18, 1972 and the Oil Pollution Act of 1990*, delegates to the EPA and Coast Guard various responsibilities assigned to the President under CWA Section 311 and the Oil Pollution Act of 1990.

Other relevant EOs include: EO 11990 (1977), *Protection of Wetlands* and EO 11988 (1977), *Floodplain Management*.

RCRA 1976, as amended by the HSWA of 1984, has the objectives to protect human health and the environment, reduce waste and conserve energy/natural resources, and to reduce or eliminate generation of hazardous waste:

- Subtitle D - solid waste (encourages states to develop and implement solid waste management plans to provide capacity).

- Subtitle C - hazardous waste program (identifies hazardous wastes and regulates their generation, transportation, and TSD; authorizes states to implement the hazardous waste program in lieu of EPA; requires permits for TSD facilities).

- Subpart S - Proposed Corrective Action Rule (provides procedures for implementing RCRA corrective action) (55 FR 30797, July 27, 1990 and 61 FR 19431, May 1, 1996).

- Subtitle I - UST (regulates petroleum products and hazardous substances stored in underground tanks; requires compliance with performance standards for new tanks; and requires leak detection, prevention, closure, financial responsibility, and corrective action).

CERCLA of 1980, as amended by the SARA of 1986 (42 U.S.C. 9601 et seq.) provides broad Federal authority to respond directly to releases or threatened releases of hazardous substances that may endanger public health or the environment. SARA defines the process Federal agencies must follow in undertaking RA, including a requirement that EPA make the final selection of remedy if there is a disagreement between the Federal agency and EPA.

The NCP (55 FR 8660, 9 March 1990) provides procedures and standards for how EPA, other Federal agencies, states, and private parties respond under CERCLA to releases of hazardous substances. The NCP authorizes the U.S. Department of Interior and other agencies, states, or entities to be the "trustees" of natural resources to recover compensatory damages for "injury to, destruction of, or loss of natural resources resulting from a discharge of oil into navigable waters or a release of a hazardous substance."

Federal Facility Compliance Act (PL-102386, October 21, 1992) directs Federal agencies to comply with Federal and state environmental laws, and provides authority to EPA to impose penalties on other Federal agencies for noncompliance. Among others, it amended Section 6001 of RCRA to waive immunity of the United States (Federal department, agency, or instrumentality of the United States) to administrative orders and civil penalties or fines associated with Federal, state, interstate, and local solid and hazardous waste management requirements. Section 3004 of RCRA was also amended to require EPA, in consultation with DOD, to identify and regulate waste military munitions which are hazardous.

1.6.2 DOD Directives.

<u>DOD Directive 5100.50 (DOD, 1973)</u>, *Protection and Enhancement of Environmental Quality*, establishes procedures and assigns responsibilities for use of DOD resources in the protection and enhancement of environmental quality and establishes the DOD Committee on Environmental Quality.

<u>DOD Directive 5030.41 (DOD, 1977a)</u>, *Oil and Hazardous Substances Pollution Prevention and Contingency Program*, sets forth DOD policy in support of the NCP.

<u>DOD Directive 4120.14 (DOD, 1977b)</u>, *Environmental Pollution, Prevention, Control, and Abatement*, implements within DOD new policies provided by EO 12088 and Office of Management and Budget (OMB) Circular A-106, and establishes policies for developing and submitting plans for improvements needed to abate air and water pollution emanating from DOD facilities.

<u>DOD Directive 6230.1 (DOD, 1978)</u>, *Safe Drinking Water*, sets forth DOD policy for provision of safe drinking water and compliance with the SDWA.

<u>DOD Directive 6050.1 (DOD, 1979)</u>, *Environmental Effects in the United States of DOD Actions*, implements the Council of Environmental Quality (CEQ) regulations and provides policies and procedures to take into account environmental considerations in DOD actions.

1.6.3 EPA Headquarters and Regional Guidance.

<u>CERCLA</u>

Guidance documents (Office of Solid Waste and Emergency Response [OSWER] Directives) for conducting various phases of a CERCLA response action have been developed or are being finalized by EPA headquarters. Key CERCLA guidance documents are identified below (also see Appendix A):

- *Guidance for Performing Preliminary Assessments Under CERCLA* (USEPA, 1991c). This document provides the PA objectives, data requirements, the procedural steps to complete the PA, and develops a site score using PA score sheets. It also provides guidelines for reviewing the site evaluation and

score, including identification of sites for emergency response actions.

- *Guidance for Performing Site Inspections Under CERCLA* (USEPA, 1992m). This document provides the approaches, data acquisition planning needs, sampling strategies, data evaluations using the SI worksheets, and reporting requirements for the CERCLA SI. The document describes the approach of using a focused SI to test the PA hypotheses, resulting in one of three recommendations: (1) site evaluation accomplished, (2) expanded SI to collect additional data, or (3) preparation of an HRS package for placement of the site on the NPL if the HRS scoring data requirements have been met.

- *Hazard Ranking System Guidance* (USEPA, 1992a) provides guidance to individuals responsible for preparing HRS packages for sites for of sites on the NPL.

- *Guidance for Conducting Remedial Investigations and Feasibility Studies Under CERCLA* (USEPA, 1988i). This guidance describes the CERCLA RI/FS process to characterize the nature and extent of contamination or risks posed by a site and to evaluate whether RA is needed. It describes the site characterization techniques, the role of a BRA, feasibility studies, and development of screening and detailed analyses of remedial alternatives.

- *Guidance for Data Useability in Risk Assessment (Part A)* (USEPA, 1992h) and *(Part B)* (USEPA, 1992k). These guidance documents provide approaches and recommendations for defining, planning, and assessing analytical data for the BRA.

- RAGS was published in two volumes: *Volume I, Human Health Evaluation Manual* (USEPA, 1989j), and *Volume II, Environmental Evaluation Manual* (USEPA, 1989b). A compendium method handbook (USEPA, 1989c) was published concurrently with the Environmental Evaluation Manual. As the science of ecological risk assessment has developed, additional guidance has been published to superceed the Environmental Evaluation Manual. *Ecological Risk Assessment Guidance for Superfund: Process for Designing and Conducting Ecological Risk Assessments* was published as Interim Final on June

5, 1997 (USEPA, 1997b) and the *Guidelines for Ecological Risk Assessment*, published as Final in April 1998 (USEPA, 1998b). Volume I has four parts:

- Part A (USEPA, 1989j) provides a detailed discussion on how a BRA should be conducted. It presents key components of a risk assessment: data collection and evaluation, exposure assessment, toxicity assessment, risk characterization, and uncertainty discussion.

- Part B (USEPA, 1991d) presents the methodologies and algorithms to calculate risk-based PRGs for individual chemicals in the soil, ground water, and air media, and the transformation of PRGs to RGs or cleanup levels using site-specific information. It stresses that risk-based cleanup levels are to be considered along with ARARs, remediation technology, and analytical detection limits (DLs), etc., in the risk management and remedy selection processes.

- Part C (USEPA, 1991e) presents the approach and risk information used to evaluate remedial alternatives during the FS. The evaluation (either qualitative or quantitative) compares risk-based benefits of alternatives, investigates potential risks to the nearby communities (short-term and long-term/residual) and remediation workers (short-term), determines the need for engineering controls to mitigate potential risks, and assesses the need for a 5-year review indicated in the NCP. The guidance describes selected remediation technologies and provides references for quantifying the potential releases from conducting such remedial activities.

- Part D (USEPA, 1998a). The EPA was directed to establish national criteria to plan, report, and review Superfund risk assessments. The RAGS Part D approach includes three basic elements: (1) Use of the Standard Tools, (2) Continuous Involvement of EPA Risk Assessor, and (3) Electronic Data transfer to the National Superfund Database. Additionally, EPA is developing standard

approaches for lead risks, radionuclide risks, probabilistic analyses, and ecological evaluation that will be issued as revisions to RAGS Part D.

The approach contained in RAGS Part D is intended for all CERCLA risk assessments. Its use is also encouraged in ongoing risk assessments to the extent it can efficiently be incorporated into the risk assessment process. Part D is also recommended for non-NPL sites, BRAC sites and RCRA sites when appropriate. Chapter 1 of RAGS Part D provides more detailed guidelines regarding its applicability as a function of site lead and site type. Each EPA region will determine the site-specific applicability, but USACE risk assessors should consider its use on all projects.

- EPA regional guidance documents for risk assessment. Various EPA regions have also supplemented the national EPA risk assessment guidance with their own policies and procedures for use in conducting a BRA. These guidance documents, in the form of memoranda, directives, or stand-alone documents, address a wide range of issues. These issues include adjustment of critical toxicity factors, data presentation and qualifications, use of MC simulations in risk characterization, selection of ground water data to estimate the reasonable maximum exposure point concentration, toxicity equivalency factors for polycyclic aromatic hydrocarbons (PAHs), soil/dermal adherence factors, midrange (CT) values for exposure parameters, selection of COPCs, screening risk assessment methods, and others.

RCRA

Limited guidance has been developed for conducting various phases of a RCRA facility response action to address current or past releases. The key RCRA guidance documents that are available are identified below:

- *RCRA Facility Assessment Guidance* (USEPA, 1986) provides guidance for conducting facility assessments to reflect developments of the RCRA corrective action programs. Also clarifies the definition of SWMU.

- *RCRA Corrective Action Interim Measures Guidance* (USEPA, 1988g) assists EPA regions and states in performing corrective action interim measures to mitigate or remove an exposure threat presented by releases.

- *RCRA Corrective Action Plan* (USEPA, 1988a) provides technical framework for developing corrective action orders and corrective action permit requirements.

- *RCRA Facility Investigation (RFI) Guidance* (USEPA, 1989f) provides general guidelines for performing health and environmental evaluations are described in this four-volume guidance manual. With regard to performing environmental risk assessments, this guidance is substantively equivalent to RAGS and references the CERCLA methodology.

1.6.4   State Requirements/Guidance. HTRW risk assessors and PMs need to be aware of any risk assessment procedures, data needs, or programs specific to the state in which their site is located. Almost all states have been authorized for RCRA permitting; some have corrective action authorities. Many states have statutes and regulations that address uncontrolled hazardous waste sites and SWMUs associated with regulated RCRA facilities. Also, many states have primacy in the water pollution control program (under the CWA) and have either adopted EPA criteria or developed their own water quality standards. Many states have adopted the use of risk assessment for corrective action to demonstrate "how clean is clean," to develop site-specific cleanup goals, to evaluate facilities burning hazardous waste, or for other uses.

Some states (e.g., California and New York) have risk assessment policies which may be interpreted as substantially similar to RAGS. Other states (e.g., Connecticut and Kentucky) have adopted RAGS as a matter of policy. Some states (e.g., Ohio and Massachusetts) have developed formal risk assessment guidelines, ranging from calculation of exposure point or background concentrations to the adjustment of critical toxicity values. Ohio and Tennessee recommend a health risk assessment be performed for RCRA corrective action and closure to demonstrate "how clean is clean." Some states (e.g., Kentucky, Michigan, New

York, Oregon, and Texas) allow the use of risk assessment to derive ACLs and medium-specific action levels or risk reduction standards. A few states (e.g., Connecticut and Illinois) have simple procedures in place (such as 20 times the maximum concentration of contaminants for the toxicity characteristics or use of equilibrium partitioning) to derive preliminary soil/sediment cleanup levels. In general, risk assessment or analysis procedures vary from state to state, and sometimes within different departments or among state agencies.

1.6.5   Others.

U.S. Army (USA)

AR 200-1 (USA) designates USACHPPM (formerly the U.S. Army Environmental Hygiene Agency) to oversee and recommend approval or disapproval on behalf of the U.S. Army Office of The Surgeon General on all risk assessments prepared by executing agencies for Army IRP sites, Army BRAC sites, and FUDS. USACHPPM is the DOD Lead Agent and Army liaison office for the ATSDR program.   USACHPPM works with the military components and ATSDR to prevent exposures at hazardous waste sites and to prevent any potential adverse health effects associated with such exposures.   USACHPPM executes the Memorandum of Understanding between DOD and ATSDR, and identifies requirements and negotiates and Annual Plan of Work with ATSDR.

U.S. Air Force (USAF)

The Office of the Air Force Surgeon General's Biomedical Engineering Service (BES) is responsible for providing technical support for all Air Force DERP CERCLA activities. The Air Force *Installation Restoration Program Management Guidance* (USAF, 1989) and Fiscal Year (FY) 93/94/95 *DERA Eligibility and Programming Guidance* (USAF, 1992) provide guidance in this area. Work relating to hazardous waste management activities under RCRA is performed by the BES in accordance with Air Force Regulation 19-7 and *USAF Hazardous Waste Management Policy* (USAF, 1991). Currently, the environmental service centers for USAF, such as the Air Force Center for Environmental Excellence, USACE, or the risk assessors at respective

Major Air Force Commands review risk assessments in coordination with the Air Force Surgeon General.

U.S. Navy and Marine Corps

The Chief of Naval Operations directive OPNAVINST 5090.1B (DON, 1994), Department of the Navy (DON), assigns command responsibilities and provides Navy policy to comply with environmental laws and regulations. *The Navy and Marine Corps Installation Restoration (IR) Program Manual* (DON, 1992) describes the Navy organization/responsibilities in support of IRP, priority for funding, research, training, and reporting requirements including preparation of Pollution Control Report to satisfy the OMB Circular A-106 reports to EPA. The Naval Environmental Health Center, under the direction of the Bureau of Medicine and Surgery, provides a wide range of medical consultative services to the Naval Facilities Engineering Command community in support of the IRP, the BRAC Program and other related environmental projects. Consultative support services include but are not limited to review of IRP and BRAC program documents (e.g., work plans, sampling and analysis plans (SAPs), quality assurance/quality control (QA/QC) plans; RI/FSs, risk assessments, health and safety plans) from a risk assessment and public health perspective; conducting risk evaluations or quantitative risk assessments; training in risk assessment, public health assessment, health and safety plans, and risk communication; sponsoring the 3-day tri-service Environmental Risk Communication and Public Dialogue Workshop; negotiating with regulators regarding the use of realistic exposure assumptions; assisting in developing community relations plans; assisting in establishing Restoration Advisory Boards; assisting in preparing correspondence from a risk communication perspective; preparing posters for public exhibits and public meetings; acting as the DON liaison for ATSDR issues.

USEPA

The USEPA has published a number of enforcement policies and procedures for Federal facilities, e.g., *Federal Facilities Compliance Strategy* (USEPA, 1988j), *Enforcement Actions Under RCRA and CERCLA at Federal Facilities* (USEPA, 1988b), *Evaluation Process for Achieving Federal Facility Compliance* (USEPA, 1988c), *Federal Facilities Negotiations Policy*

(USEPA, 1989h), and *Federal Facilities Hazardous Waste Compliance Manual* (USEPA, 1990a). All Federal agencies are required to comply with hazards waste regulations and the NCP in the same manner as the private sector.

U.S. Department of Energy (DOE)

The DOE has issued a number of orders (5400 series and others) addressing a variety of environmental statutes and requiring all facilities to comply with the applicable environmental laws and regulations. For example, DOE Order 5400.2A (DOE, 1993) sets forth policy, direction, and procedures for coordinating environmental compliance issues and DOE Order 5400.4 (DOE, 1989) addresses "CERCLA Requirements." The Office of Environmental Guidance of DOE has a plan in place to develop a comprehensive guidance and training program for its field facility staff and Environmental Restoration Project Managers. In the area of risk assessment, the DOE guidance or information briefs include: *Integrated Risk Information System* (DOE, 1991), *CERCLA Baseline Risk Assessment* (DOE, 1992a), and *Use of Institutional Control in CERCLA Baseline Risk Assessment* (DOE, 1992b).

1.7 FEDERAL FACILITY AGREEMENT

Although there may be subtle differences between an FFA and an IAG, these terms are used interchangeably under CERCLA Section 120 which addresses both NPL and non-NPL sites. This section focuses on the need for early planning and negotiation of an FFA among the USACE customer (a Federal agency), EPA, and the state agency (as appropriate). To accomplish this objective, the HTRW project team member (i.e., the risk assessor) and others should work cooperatively to develop statements/languages or addenda to the FFA early in the HTRW project cycle to define a flexible framework or process for RMDM and to facilitate site closeout protective of human health and the environment.

EO 12580 delegates DOD to conduct response action under Section 104 of CERCLA (as amended by SARA) to address releases on DOD facilities or originating from the facilities. The order requires that the response action be conducted in accordance with Section 120 of CERCLA. According to CERCLA Section 120(e)(1), DOD is directed to enter into an IAG with EPA for RA within 180 days of EPA's review of the RI/FS. In the *Federal Facilities Hazardous Waste Compliance Manual* (USEPA, 1990a), EPA states, "At a minimum, the IAG must include a review of cleanup alternatives considered and the remedy selected, a schedule for cleanup accomplishment, and arrangements for operation and maintenance."

To address non-compliance issues at a Federal facility (e.g., a DOD installation), EPA may issue a complaint known as Notice of Noncompliance (NON). After such an issuance, EPA and the Federal facility enter into negotiation for a Federal Facility Compliance Agreement which resolves compliance violations and stipulates agreed-upon remedy, compliance schedule, and reporting and record keeping requirements. The target date for concluding such an agreement is within 120 days from the date of NON issuance (USEPA, 1990a). Since RCRA corrective actions are generally required at the time of RCRA Part B permitting or permit renewal, the Federal facility may be issued a RCRA Section 3008(h) corrective action order rather than a NON.

"Executive branch disputes of a legal nature are properly resolved by the President or his or her delegate..." (USEPA, 1990a). In view of the above, and for the purpose of this handbook, the risk assessor should provide assistance to the USACE's PM, risk manager, and the USACE customer so that an FFA or IAG can be successfully negotiated to provide a framework for RMDM and to initiate actions to protect human health and the environment where these actions are needed. The risk assessor and the HTRW project team may consider the following areas for assistance to be provided to the USACE customer concerning the FFA negotiation; these areas have been identified in the DOD-EPA Model IAG Language (USEPA, 1989h):

1.7.1   Basis for Interim Remedial Action (IRA) Alternatives. For purposes of this guidance, IRA may be interpreted as interim corrective measure under RCRA or interim removal action under CERCLA. One purpose of the FFA is to identify IRA alternatives which are appropriate at the site prior to the implementation of final RA(s). To identify such alternatives, the exposure area, the exposure pathways which contribute to the principal threat at the site, and the receptors/resources must also be identified. For the purpose of the FFA, a

statement may be entered which indicates the basis for identifying IRA alternatives. This statement should address the following:

- The approach for conducting a screening risk analysis of the Exposure Units (EUs) (USEPA, 1991a), SWMUs, or the AOCs.

- The evaluation method for the risk assessment/analysis results (qualitative or quantitative).

- RMDM considerations (see Chapter 6) for identifying and/or selecting the IRA alternatives.

1.7.2   Requirements for RI/RFI and FS/CMS. Another purpose of the FFA is to provide a framework for investigating, assessing the impact, and evaluating remedial options to protect public health and the environment. Such a framework, consistent with the NCP and the RI/FS guidance (USEPA, 1988i), may be modified and formally incorporated in the FFA to meet the site-specific and project requirements. Statements or languages or addenda to the FFA may be prepared by the risk assessor and the project team to serve as a basis for determining the extent of data collection, data evaluation, assessment of baseline risk, and evaluation of remedial alternatives. The HTRW TPP process (USACE, 1998) and associated DQOs should be identified as the framework for determining data needs, data use, and quality. The point of departure for NFA and/or monitoring only based on acceptable carcinogenic risk or hazard should be identified in the FFA (USEPA, 1991a). The statement should indicate the need for evaluating uncertainties in risk assessment by the use of multiple descriptors (i.e., RME, CT, population, and individual risks). One important statement that should also be considered for complex sites is the need for a probabilistic risk assessment to identify the confidence level of unacceptable risk or hazard, when the point estimate of risk derived by the deterministic approach (e.g., RAGS Part A, USEPA, 1989j) has marginally exceeded the acceptable risk or hazard levels. These probabilistic risks (cumulative function distribution) should be identified as an input into the RMDM for these site actions.

1.7.3   Expedited Cleanup Process. Both DOD and EPA are in agreement that early action or accelerated cleanup may be needed to stabilize the site and to facilitate implementation of the final remedies. However, the basis for such action is not well defined, except that the actions are intended to control contaminant migration, to reduce exposure, and to accelerate response. In addition to time-critical and emergency response actions where safety and acute hazards are involved, the risk assessor and the project team can provide valuable input to the USACE customer and risk manager for such expedited actions. This can be rather quickly accomplished by comparing the measured media concentrations with available human health and ecological risk-based protective criteria. This may be useful for relatively straight-forward sites, such as drum removal, product removal, and containment. For response actions at a complex site, a BRA may be more appropriate, however, and expedited cleanup would not be done. All decision criteria for eliciting response actions to protect environmental components should be well thought out, reasonable, and consistent with current EPA guidance.

1.7.4   Units Excluded from the Agreement. RCRA and CERCLA integration issues should be addressed in the FFA in unambiguous terms. This is particularly true for sites of which the state agency is also an interested party or natural resource trustee in the agreement. Some state agencies have their own risk assessment policies and guidance, and RMDM criteria which may vary substantially from those of EPA (EPA's procedures under RCRA and CERCLA are judged to be substantially equivalent at this time). The risk assessor should review state policies, guidance, and requirements, to identify any critical risk assessment/risk management issues for the PM and the customer for resolution. These issues should be addressed and resolved in the FFA negotiations. If not successful, separate FFAs may be needed to address RCRA and CERCLA units within the facility. The USACE and customer's legal counsels should be contacted for briefing on these issues early in the process.

CHAPTER 2

2.0 PLANNING FOR AN HHRA

2.1 INTRODUCTION. The consistent standardized approach presented in this guidance document was devised to assure consistent treatment among sites. Numerous other resource materials, guidance documents, bulletins, memoranda, technical manuals, and books that address the general HHRA approach and scoping of site-specific data needs are available from EPA, other regulatory agencies, and scientific sources. A number of these resources are referenced in Appendix A. The generally accepted approach to performance of an HHRA is presented in RAGS (USEPA, 1989j), and a thorough understanding of the process is prerequisite to working within the USACE program. This guidance will not reiterate RAGS, but the following paragraphs will provide the USACE risk assessor and risk manager with the details necessary to focus investigations toward site closeout and to provide USACE policies and procedures on the HHRA process, along with "how to" and "where to find" knowledge for evaluating the scope, design, and conduct of a site-specific HHRA.

2.1.1 Purpose of the HHRA. The HHRA is an integral component of the PA/SI, RI/FS, RD/RA[11], and emergency response processes, serving multiple functions in decision-making:

- The HHRA provides an evaluation of the potential human health risks under baseline (i.e., no action) conditions.

- The HHRA helps determine the need for RA at the site.

- The HHRA provides a basis for determining RGs for chemicals in site media.

---

[11] As stated previously, this document assumes the processes involved in CERCLA and RCRA investigations to be equivalent. For the rest of these discussions, CERCLA terms only will be used. It may be assumed that the procedures are also appropriate for the equivalent RCRA phase.

- The HHRA provides a basis for comparing different remedial alternatives.

- The HHRA provides a consistent and widely accepted methodology for assessing potential health risks, allowing for comparison of potential health risks between sites.

2.1.2 Objectives of the HHRA. The goal of the HHRA is to provide the necessary information to assist risk managers in making informed decisions. The HHRA provides important risk management input at various project phases, identifying receptors or resources to be protected, as well as limitations and uncertainty.

The HHRA should provide an objective, technical evaluation of the potential impacts posed by a site, with the risk characterization clearly presented and separate from any risk management considerations. Although risk assessment and risk management are separate activities, the risk assessor and risk manager need to work together at various stages throughout the project to define decision data needs. In the HHRA, the risk assessor needs to present scientific information in a clear, concise, and unbiased manner without considering how the scientific analysis might influence the regulatory or site-specific decision. The risk assessor is charged with:

- Generating a credible, objective, realistic, and scientifically balanced analysis.

- Presenting information on the problem, effects, exposure, and risk.

- Explaining confidence in each assessment by clearly delineating strengths, uncertainties (as well as an estimation of the effects of the uncertainties, both magnitude and direction), and assumptions, along with impacts of these factors (USEPA, 1995c).

The risk assessor does not make decisions on the acceptability of any risk level for protecting the receptors or selecting procedures for reducing risk. The HHRA is used by the risk manager, in conjunction with regulatory and policy considerations, to determine the appropriate response actions at the site.

2.1.3 Minimum Requirements. The provision for "minimum requirements" for the HHRA is an important concept. The risk assessor should identify particular minimum requirements for activities preceding and used in the HHRA to assure that critical factors are addressed. Early in the process of planning the HHRA, the risk assessor should also confer with the end users of the assessment to identify all factors that need to be addressed by the HHRA. The HHRA should be developed with its end uses in mind. Early interaction with risk managers and remedial designers is needed to obtain information on the risk management options likely to be considered if RA is required. This is not to infer that the HHRA should be "tailored" to specific remedial options, for that would compromise the objective nature of the assessment. However, if the risk manager or remedial designer needs certain information (for example, what depth of soil should be considered surface soils, given projected site use or exposure during remediation), the HHRA should provide the basis that will allow this question to be answered (within the appropriate boundaries of the HHRA).

2.1.4 Technical Requirements. The technical requirements of the HHRA should be considered early in the site planning and investigative phase to assure that appropriate information is gathered. It is important that the risk assessor be involved in the early planning stages of field investigations to develop the CSM, which will help guide the identification of site media to be sampled, and to assist in designing the chemical analytical scheme. The risk assessor should also assist in DQO development for performance-based methodology, design of the data review process, and performance of the data useability assessment. This will help assure that the best possible and most relevant data are available for use in the HHRA.

2.1.5 Technical Basis. Risk assessments developed for the various activities will have slightly different requirements, require a different scope, and will involve a different level of effort. However, the technical basis for performing the risk assessment is essentially the same. The main description of the risk assessment methodology is provided below, and discussions of all types of risk assessments are based upon this model. Therefore, the information presented is necessary to the understanding of other risk assessment applications. Each type of risk assessment is discussed in subsequent chapters.

The HHRA is one component of overall site investigation and remedial activities. It should be developed with a recognition of how it is supported by preceding and concurrent components of site activities, such as sampling and analysis for the ERA effort, and how it supports and shapes the subsequent components, such as RD. Although the HHRA is performed to achieve several specific objectives (describing current and future human health risks), it needs to be coordinated with other site activities (e.g., ERA) and needs to be responsive to other general site concerns (e.g., restoration, mitigation, litigation) and the resources (cost and schedule to be met) available.

The risk assessment process has been separated by convention into four subdisciplines: hazard identification, dose-response assessment, exposure assessment, and risk characterization (NRC, 1983 and NRC, 1994). Hazard identification is the process of determining whether exposure to an agent could cause an increase in the incidence of adverse health effects. The dose-response assessment evaluates the relationship between the dose of an agent and the probability of producing adverse effects. Exposure assessment evaluates the combination of chemical uptake and potential routes of exposure. Finally, risk characterization summarizes and interprets the information and evaluates the limitations and uncertainties in the risk estimates (NRC, 1994).

Risk assessments have different applications in different regulatory programs. This document discusses the application of risk assessment in the following phases of site activity:

- PA/SI.

- RI.

- FS activities, including development of remediation levels and comparative risk assessments associated with selected remedial options, followed by the evaluation of short term risks associated with the implementation of the selected remedial option.

- RD/RA activities, including potential need to further evaluate short-term risks for the purpose of designing/ implementing control measures.

- Assessment of residual risk after implementation of the selected remedial option.

2.1.6 Planning and Problem Identification.
Planning and problem identification are critical to the success of the HHRA and its usefulness with respect to remediation planning. To assure that the scope of the HHRA is sufficient for making risk management decisions, the risk assessor must always be mindful of the question, "Do the data and approach support RMDM?"

In identifying data needs for the HHRA, the risk assessor must fully understand the customer goals and the regulatory program(s) driving the HTRW project execution. The concept of TPP is fully explained in EM 200-1-2 (USACE), which emphasizes the need for the data users (e.g., the risk assessor) to identify minimum data requirements for the tasks to be performed.[12] The concept of "minimum requirements" for the HHRA is important in that it identifies certain aspects for data collection activities preceding the risk assessment to assure that critical data gaps or factors are addressed.

The approaches and contents of the anticipated risk assessment should be explained or discussed in the project planning stage in unambiguous terms. An iterative, tiered approach to the risk assessment, beginning with screening techniques, is used to determine if a more comprehensive assessment is necessary. The nature of the HHRA depends on available information, the regulatory application of the risk information, and the resources available to perform the risk assessment. Informed use of reliable scientific information from many different sources is the central

feature of the process (USEPA, 1995a,c). The TPP process should produce an outline for a site-specific HHRA that is credible, objective, realistic, and scientifically-balanced.

Throughout the planning discussions, the risk assessor should strive to point out potential setbacks, problems, or difficulties that may be encountered in a "real world" situation. When special circumstances (e.g., lack of data, extremely complex situations, resource limitations, statutory deadlines) preclude a full assessment, such circumstances should be explained and their impact on the risk assessment discussed. The risk assessor should also explain the minimum data quality considered to be acceptable, how non-detects will be treated, and how medium-specific data will be evaluated or compiled to derive or model the exposure point concentration in the risk assessment.[13]

2.2 PLANNING CONSIDERATIONS

2.2.1 Coordinating HHRA and ERA Planning.
Planning for a HHRA should be conducted concurrently with that for an ERA in that these two efforts often have similar data needs. Data needs for the ERA, however, eventually focus on developing remedial alternatives that are protective of ecosystem components, while the HHRA focuses on developing remedial alternatives that are protective of a single species, humans.

Coordinated planning efforts for the HHRA and ERA efforts, particularly where there is to be an expedited cleanup, should include consideration of the following:

- Overlaps in information needs with regard to human and ecological food chain issues.

- Benefits of the cleanup and the effectiveness of presumptive remedies.

---

[12] The HTRW TPP process is a four-phased (Phase I through Phase IV) process that begins with the development of a site strategy and ends with the selection of data collection options. Throughout the process, USACE HTRW personnel of various disciplines and responsibilities (some of whom may assume multiple responsibilities) work closely together to identify data needs, develop data collection strategy, and propose data collection options for the customer. The HTRW data quality design process implements the EPA's DQO process, which is an iterative process applicable to all phases of the project life cycle.

[13] For example, if the RI data are skewed, it may be necessary to address site risk by evaluating hot spots separately. The risk assessor may wish to indicate this in the Work Plan, in order to characterize hot spot areas without delaying the assessment of risks for the non hot-spot areas.

- Ecological impacts from removal or remedial activities designed to protect human health.

- Identification of hot spots that may impact both human health and ecological receptors.

- Identification of the key assumptions and criteria common to the HHRA and ERA that may drive cleanup decisions and focus the decision making process.

- Identification of areas of greatest concern that may be addressed early as discrete tasks, thereby allowing priority to be given to those (removal/remedial) actions that achieve the greatest protection of the environment and human health for the capital (dollars) spent.

- Activities common to both the human health and ecological risk efforts that support DOD responsibilities as a Natural Resource Trustee or help coordinate between multiple Natural Resource Trustees where jurisdictions or responsibilities overlap.

2.2.2   Coordination   with   Natural   Resource Trustees. In the risk planning process, on Superfund sites in particular, it is also important for the risk assessor, risk managers, the technical team, and decision makers to coordinate with natural resource trustees (e.g., DOD, the state, the National Oceanic and Atmospheric Administration [NOAA[14]], the U.S. Fish and Wildlife

---

[14]   NOAA's Coastal Resource Coordination Branch (CRCB) works with EPA through all phases of the formal remedial process at Superfund waste sites. The CRCB acts for the Dept. of Commerce as trustee for natural resources such as anadromous and marine fish. Coastal Resource Coordinators (CRCs) and an advisory staff of environmental, marine, and fisheries biologists provide technical support and expertise to EPA, DOD, and other agencies during response and cleanup at coastal waste sites.   The CRCs and supporting staff recommend appropriate environmental sampling, coordinate with other natural resource trustee agencies to build consensus on natural resource issues, and recommend appropriate clean-up levels. The CRCB works with EPA to gain cost-effective remedies that

Service, the U.S. Forest Service, and the Bureau of Land Management) at the earliest possible stage. In this way, the trustee can be assured that potential environmental concerns are addressed, and conclusion of action may be expedited (USEPA, 1989g, 1989h, and 1989i). Coordination with natural resource trustee agencies such as NOAA provides for the exchange of ideas and issues to assure the technical adequacy of the RI/FS, to assure the protectiveness of the selected remedy for trust resources, and to provide for proper restoration and mitigation for injured resources. Coordination also allows DOD access to the trustees' specific skills, information, and experience.   This interaction may occur through a variety of informal and formal forums, including but not limited to: preliminary scoping and drafting of work plans, review of final work plans and subsequent data, technical review committees, PM meetings, and public information meetings.

2.2.3   RAGS, Part D: Standardized Planning, Reporting, and Review of Superfund Risk Assessments. EPA Administrator, Carol   Browner, called for an improvement in the transparency, clarity, consistency, and reasonableness of risk assessments (USEPA, 1995c). Subsequently, the October 1995 Superfund Administrative Reform #6A directed EPA to establish national criteria to plan report & review Superfund risk assessments. As a result, the *Risk Assessment Guidance for Superfund (RAGS): Volume I: Human Health Evaluation Manual; Part D, Standardized Planning, Reporting, and Review of Superfund Risk Assessments* (USEPA, 1998a) was developed.   Additionally, EPA is developing standard approaches for lead risks, radionuclide risks, probabilistic analyses, and ecological evaluation that will be issued as revisions to RAGS Part D.

The RAGS Part D approach includes three basic elements: (1) Use of Standard Tools, (2) Continuous Involvement of EPA Risk Assessor, and (3) Electronic Data transfer to National Superfund Database. Brief descriptions of the three components follow:

2.2.3.1 Use of Standard Tools. The Standard Tools include a Technical Approach for Risk Assessment

---

minimize residual resource injury without resorting to litigation. CRCs are in most EPA regions.

(TARA), Standard Tables, and Instructions for the Standard Tables. The TARA is a "road map" for incorporating continuous involvement of the EPA risk assessor throughout the CERCLA remedial process for a particular site. The TARA should be customized for each site-specific HHRA as appropriate. Electronic templates for the Standard Tables have been developed in Lotus and Excel for ease of use by risk assessors. For each site-specific risk assessment, EPA recommends the Standard Tables, related Worksheets, and supporting information first be prepared as Interim Deliverables for EPA risk assessor review, and should later be included in the Draft and Final BRAs.

Instructions for the Standard Tables have been prepared corresponding to each row and column on each Standard Table. The Instructions should be used to complete and/or review Standard Tables for each site-specific HHRA. Instructions, example tables, and blank tables are available for download at:
http://www.epa.gov/superfund/oerr/techres/ragsd/ ragsd.html.

2.2.3.2 Continuous Involvement of EPA Risk Assessors. In this part of the document, the RPMs are instructed to use the EPA risk assessors for all CERCLA sites, from scoping through completion and periodic review of the RA. It is stated that early and continuous involvement by the EPA risk assessors should include scoping, work plan review, and site-specific customization of the TARA for each site to identify all risk-related requirements. It is also emphasized that EPA risk assessors support reasonable and consistent risk analysis and risk-based decision making.

2.2.3.3 Electronic Data Transfer to a National Superfund Database. Summary-level site-specific risk information will be stored in a National Superfund Database (CERCLIS 3) to provide data access and data management capabilities to all EPA staff. These risk-related summary data represent a subset of the data presented in the Standard Tables. The electronic versions of the Standard Tables (Lotus and Excel) are structured to be compatible with CERCLIS 3.

2.2.3.4 RAGS Part D Applicability. The approach contained in RAGS, Part D is intended for all CERCLA risk assessments. Its use is also encouraged in ongoing risk assessments to the extent it can efficiently be

incorporated into the risk assessment process. Part D is also recommended for non-NPL sites, BRAC sites and RCRA sites when appropriate. Chapter 1 of RAGS Part D provides more detailed guidelines regarding the applicability of RAGS Part D as a function of site lead and site type. Each region will determine the site-specific applicability, but USACE risk assessors should consider its use on all HTRW projects.

2.2.4 The HTRW TPP Process. EM 200-1-2 (USACE) provides guidance on data collection programs and defines DQOs for HTRW sites. DQOs define the project's data needs, data use, number of samples required, the associated QA requirements (e.g., quantitation limits(QLs), blanks, split and duplicate samples, etc.), and level of confidence or acceptable data uncertainty for the requisite data. DQOs are generated at the final phase (Phase IV) of the TPP process after the customer has selected the preferred data collection program. The process includes evaluation of previously collected data, and assessment of the need for additional data to support the current or subsequent phases of the project. This coordinated TPP effort is designed to satisfy the customer goals, applicable regulatory requirements, and minimum technical data requirements for performing site investigations.

Throughout the process, USACE HTRW personnel of various disciplines and responsibilities work closely together to identify data needs, develop data collection strategy, and propose data collection options. The HTRW TPP process is consistent with the EPA's 7-Step DQO process, which is an iterative process applicable to all phases of the project life cycle. The DQO development process is considered to be a Total Quality Management tool (USEPA, 1989e). This is key to assuring successful planning and performance of the risk assessment.

Phases I through IV (described below) of the TPP process address site investigations methodically and should be incorporated throughout the entire HTRW project life cycle. Using this TPP process, the risk assessor will be able to define minimum information requirements for risk evaluations in support of site decisions.

2.2.4.1 Phase I - Develop Project Strategy. This phase of the TPP process involves identifying site decisions requirements and developing an approach to address these requirements. Site strategy is broadly defined in the beginning of a project at this stage. As the project progresses into subsequent phases, the strategy is refined based on an improved understanding of the site. The risk assessor is crucial to the development of appropriate site strategy in this phase and the identification of data needs and the associated quality requirements to support risk management decisions. In this planning phase, site conditions are reviewed qualitatively, and a preliminary CSM is developed to help define the study elements for the current and subsequent TPP phases.

2.2.4.2 Phase II - Identify Potential Data Needs. This phase of the TPP process focuses on identifying data needs and minimum data quality requirements to support site decisions. Data users identify potential data needs and their respective proposed QA/QC requirements based on site background, regulatory information, and the customer's goal. At this phase, the compliance, remedy, and responsibility data users, who have specific data needs, present their data requirements along with the data needs identified by the risk assessor. The objective is to identify the data needs and quality requirements of all project team members.

2.2.4.3 Phase III - Identify Data Collection Options. This phase of the TPP process incorporates previously identified data needs and project constraints in designing a data acquisition approach. Various sampling approaches can be used, ranging from purposive (judgmental or biased) to representative (random) sampling methods. Additionally, various analytical schemes may be used such as screening or definitive data. This phase of TPP also involves identifying the optimum sampling/data collection scheme so as to minimize mobilization, field sampling, and demobilization efforts and costs. The objective of Phase III is to identify options (preferably two or three options, out of which one is an optimum option) for presentation to the customer in Phase IV.

2.2.4.4 Phase IV - Select Data Collection Options and Assign DQOs. This is the most important phase of the TPP process because this is where the data collection option is selected. To properly execute Phase IV, the proposed options should be clearly explained and characterized. The discussion should include data uncertainties, cost/benefits, schedule, and other constraints.

The product of this phase of the TPP process is the Statement/Scope of Work (SOW) for USACE work acquisition (either internal or the architectural-engineering contractor), a detailed cost estimate (or Independent Government Estimate) for the selected option, and DQOs for the data collection program. The DQOs explain the objectives of the data gathering activity, the data type/location, data collection and analytical scheme, the required QLs, rationale for requiring certain data quantity and quality, and how the data are to be used in making site decisions. Caution should be taken at this point about the integration and coordination between the HHRA and ERA as to how they influence DQOs. ERAs may require lower media-specific QLs than HHRAs for certain COPCs (Contaminants of Potential Ecological Concern for ERAs). The ultimate DQOs should be the lower of either for dual purpose samples, or the appropriate concentration for specific purpose samples.

2.3 ESTABLISHING THE LEVEL OF EFFORT

An important part of planning for a HHRA is determining the appropriate level of effort necessary to provide the required information. As sites will vary in complexity, so will the HHRA. Some of the site-specific factors affecting the level of effort include the following:

- The number and identity of the chemicals present.

- ARARs, to-be-considered (TBC) criteria, and applicable toxicity data.

- Reasonable future site use.

- The number and complexity of complete exposure pathways and the need for fate and transport modeling to establish exposure point concentrations.

- The required QLs based on screening values and the receptor populations.

- Quality and quantity of existing analytical data.

The following sections present requirements for planning risk assessment scopes of work for the various phases of response. In addition to the evaluation of human health risks, evaluation of the potential risks to ecological receptors should be considered as well during the planning process, as duplication of effort needs to be avoided. See the companion to this manual, EM 200-1-4, *Risk Assessment Handbook, Volume II: Environmental Evaluation* (USACE) for considerations necessary for scoping an ERA. The following discussions will help guide your data needs assessment but are not intended to be all-encompassing. Data needs depend on the complexity of the site, amount of useable data already in existence, and site-specific receptors.

2.3.1 Preliminary Risk Screening; PA/SI. This section focuses on data needs for the preliminary risk screening in the site evaluation (site assessment) phase in CERCLA and RCRA. Other HTRW site assessments, although not specifically covered under these statutes, are expected to be functionally equivalent.

2.3.1.1 Review of Existing Site Information. Before the data needs for the PA/SI are conceptualized, the risk assessor should carefully review all site background information. The data quality used to produce reports or for proposed placement on the NPL (if available) should be evaluated for this phase of execution, along with a determination of whether additional data are needed. This phase of investigation usually has little existing quantitative information available. The purpose of this review is to gain a good understanding on the following issues:

- Regulatory concerns or site problems relating to human health to aid in preliminary identification of significant exposure pathways (source, migration/transport mechanism, exposure routes, and receptors).[15]

- Physical characteristics and demographics of the site which may help define possible pathways of exposure.

---

[15] In addition to the regulatory actions or concerns, the risk assessor should also review any draft or final public health advisories, e.g., the ATSDR health consultations/advisories, state health/conservation advisories on indigenous food sources, etc. The data may be needed to accept or reject such advisories or concerns. USACHPPM should be consulted on all these public health matters.

- Operational history with regard to site waste types, probability of occurrence, and location of source areas.

This information will be valuable to begin to conceptualize possible pathways of exposure and in determining data needs to support the risk screening analysis.

2.3.1.2 CSM.

2.3.1.2.1 Data needed for the risk screening analysis should be based on a preliminary CSM which is developed in the absence of extensive site information. If there are data available from a previous study, they should be evaluated for useability in the risk screening, prior to defining additional or supplemental data needs required in the PA/SI. The CSM helps identify and visually organize potential exposure pathways and receptors and identifies those pathways which could be complete (significant or insignificant) or incomplete, for the purpose of the data needs determination. The elements of a CSM are:

- Source of contamination (ground water, surface water, soil/sediment, and air).

- Potential release mechanism.

- Migration pathways.

- Potential receptors.

- Major exposure routes (e.g., ingestion, inhalation, dermal contact).

2.3.1.2.2 The risk assessor should begin to conceptualize the data needs associated with each of the aspects of the CSM that would support the screening risk evaluation. For example, it may be determined that limited judgmental sampling data can be used to conservatively define source concentrations for direct contact exposure point concentrations. A limited number of monitoring wells may be sufficient to evaluate the ingestion route for ground water. Additionally, the physical characteristics as well as the demographics of the site are also helpful in the evaluation of potential receptors and therefore complete pathways to be evaluated in the risk screening analysis. All parts of the CSM must be examined to ascertain that each element of potentially complete exposure pathways has existing data that adequately support each component of the risk screening analysis.

2.3.1.2.3   Examples of general chemical data needs according to source/route/receptor for use in assessing potential exposure pathways for the risk screening are:

• Surface soil (incidental ingestion/dermal contact and inhalation of volatile organic compounds [VOCs] and airborne particles).

• Surface water (incidental ingestion/dermal contact).

• Ground water (ingestion, dermal contact, and inhalation of volatilized ground water contaminants due to indoor use of ground water).

2.3.1.3   Identification of Data Gaps. Once all existing data has been evaluated relative to the preliminary CSM, the risk assessor can determine what data are required to assure that the subsequent investigation can evaluate risks due to all pathways identified as complete and significant. Limited sampling of media expected to be impacted by site operations can provide adequate information to eliminate a site from further study. It is important to remember that this phase of investigation does not attempt to determine nature and extent of contamination, nor to determine the magnitude of any potential risks present. The intent is to determine whether the site poses no significant risk, and may be proposed for NFA, or must be evaluated further. This aspect is further clarified in Section 2.4.1.5.

2.3.1.4   DQOs: Determining   Data   Needs   and Documentation. The level of effort is limited in this type of assessment as is the amount of data needed to support the screening.

2.3.1.4.1   In this step the general data needs defined during   conceptualization   are   formalized   as   data requirements for each media type, specifying location of sampling, depth of samples required, chemical analysis requirements and corresponding DLs and QLs (based on health-based screening levels for comparison), confidence, and in some cases number of samples. The risk assessor may   consider   a   weight-of-evidence   approach   when specifying data requirements and

subsequently evaluating the collected data to aid in making informed site decisions at this stage of the HTRW response process.   This   is   justifiable   if   a   weight-of-evidence approach   is   used   to   support   the   evaluation   and recommendation. For example, the topography, visual observations, history of spills, runoff pattern, and the analytical results of purposive sampling would be sufficient, as a whole, to support the argument whether contamination of a medium is likely or unlikely.

2.3.1.4.2   For chemical data, however, the level of confidence will be dependent on the QA/QC, sampling method, sample handling/preservation method, analysis method, and variability of the chemical concentrations in the medium that was sampled. Reference the following EMs for the requirements for the USACE chemistry program: EM   200-1-1,   *Validation   of   Analytical   Chemistry Laboratories* (USACE); EM 200-1-3, *Requirements for the   Preparation   of   Sampling   and   Analysis   Plans* (USACE); and EM 200-1-6, *Chemical Quality Assurance for HTRW Projects* (USACE)[16]. The following factors should be considered in this planning activity in order to reduce uncertainties:

• Analytical methods should be clearly stated that identify the method DL and the QL. At a minimum, the QL must be less than the action level to prove reliable detections.

• Level of QA - Depending on data use, the level of QA for PA/SI can be field screening (i.e. screening-level data) to assist identifying sampling locations, presence or absence of contaminants with some confirmational analyses, or confirmational analyses of chemical identification   and   quantification,   e.g.,   gas chromatography/mass spectrometry method (i.e., definitive data).

• QA/QC samples - Soil or sediment samples should have field duplicates, laboratory control samples, matrix spikes, and matrix spike duplicate samples. Water samples should have field duplicates. In addition, samples for the analyses of volatile and semivolatile organic chemicals should be checked for surrogate recovery. Laboratory blanks should also

---

[16] EM 200-1-1 and EM 200-1-3 are currently in revision and should be published in FY99.

be analyzed to check for the presence of potential laboratory contaminants.

- Data variability - Detection of hot spots may not be the objective of the sampling program under PA/SI. The number of samples required to represent the level of contamination with a predetermined level of confidence will depend on the uniformity or homogeneity of the contamination. This information can only be obtained via historical documentation or previous sampling events.

2.3.1.5 Risk Screening. The essence of the screening-level assessment is to determine if the site may be eliminated from further concern or requires further study, based on past releases, ARARs, and/or human health impacts. The project study elements may include current and future land use and the population characteristics, based on the evaluation of the preliminary CSM. However, this is a preliminary screening, and is intended to be a conservative assessment of potential site risks. Usually, the risk screening employs the highest detected concentrations and compares them with health-based screening levels, appropriate for the current and projected future land use of the site. Generally, exceedance of these conservative values is only an indication that further study may be required, and does not indicate that risks are significant, or that they even exist. See Chapter 6 for a complete discussion of risk management issues appropriate at this phase of investigation.

2.3.1.6 Reporting Requirements. The following elements should be clearly presented in the PA/SI Report:

- Preliminary CSM, adjusted according to any new information identified during the field investigation.

- DQOs and an evaluation of whether or not they were met.

- The comparative risk analysis (the evaluation of maximum detected values relative to health-based screening levels).

- Discussion of all uncertainties and their potential impact on the results of the risk screen.

2.3.2   HHRA; RI. The sections below focus on HTRW scoping for the baseline HHRA[17] performed in the RI. The purpose of the BRA is to estimate the degree of risk associated with the site to human receptors in order for an informed risk management decision to be made regarding future actions. Generally, if the baseline risk is acceptable, there should be little basis for the FS or RD/RA.

2.3.2.1   Review of Existing Data. At this project phase, the risk assessor should have some understanding of the site background and descriptions of site characteristics from the review of the preliminary (PA/SI) data, contained in the Federal Facility Docket or pertinent project files. This information will be useful in focusing the data needs required to prepare the BRA. Before the data needs are determined, it is recommended that the risk assessor carefully review all site background information and site assessment reports, available state and/or EPA reports, removal action information (if applicable), SI worksheets, notes, or photos, etc. These studies, reports, and photos help the risk assessor begin to focus on aspects of the site which will require evaluation in the RI under the BRA.

2.3.2.1.1 Historical data collected for purposes other than BRAs may be available from previous investigations, facility records, permit applications, or other sources. However, historical data sets may be limited by the lack of information on laboratory and QA/QC procedures, or are obtained from the wrong media and wrong location for use in the BRA. Data from historical sources may or may not be appropriate to use in the quantitative BRA and should be reviewed for useability. When evaluating historical or purposively collected data, a number of factors need to be evaluated.

2.3.2.1.2 The review focuses on the following issues:

- Regulatory concerns (or newly identified concerns) relating to specific receptors, COPCs, and the

---

[17] For the purposes of this text, Baseline HHRA and BRA can be used interchangeably. BRA will be used here to avoid confusion with established EPA guidance for HHRA (USEPA, 1989j). It is understood that the evaluation of potential environmental risks, or ERA, is an integral part of the BRA.

exposure pathways of concern, as well as those pathways exceeding health-based screening levels in the PA/SI screening-level HHRA.

- Source areas which have been identified in previous studies and the need for further quantification to evaluate extent of contamination and risks.

- Spatial relationships of pathways, and the need for segregation as EUs or OUs to properly evaluate risks to a number of receptor groups.

- Possible transport pathways and available temporal data, chemical/physical data describing degradation, attenuation, or migration of chemicals in the environment.

- All possible current site receptors, including those that may be considered sensitive, to begin grouping by classification: agricultural, residential, etc.

2.3.2.2 CSM. The CSM is the basis for development of the level of effort for the risk assessment and the DQOs that will be defined in the SOW. The CSM presents contaminant sources, release mechanisms, transport media, exposure pathways, exposure points, and receptors for current and future land uses. The CSM helps organize and identify those pathways which are complete (significant or insignificant) and incomplete. The risk assessor should review site data and information collected in the previous project phases (PA/SI) to refine the CSM. The information should be able to assist the risk assessor in developing a more definitive CSM or multiple CSMs if there are multiple OUs. A CSM for ecological receptors should be developed concurrently with the CSM. EM 200-1-4, Vol. II (USACE) describes this process. The CSM for the BRA should help define and organize by pathway:

- Classes of COPCs (information concerning the source characteristics, medium contamination, and background chemicals is needed to identify COPCs).

- Potential target media (ground water, surface water, soil/sediment, and air).

- Potential receptors exposed to the target media.

- Major exposure routes or pathways of concern (e.g., direct contact resulting in soil or sediment ingestion or dermal absorption of contaminants in the media, consumption of food chain crops or species, ground water ingestion, and inhalation of contaminants in ambient air).

- Migration and transport potential of site chemicals from the source, including the effect of existing institutional controls or removal actions (e.g., ground water capture well systems).

- Potential secondary sources of contaminants, and their release/transport mechanism(s).

2.3.2.3 PRG Development. PRGs should be prepared or obtained to assist in planning. PRG values will be used in establishment of adequate QLs for the analytical scheme. In order to characterize risks, QLs must be lower than the PRG value used. Values developed by EPA Regions such as Region 3 Risk-Based Concentrations (RBCs), or Region 9 PRGs may be used for direct comparison, or the risk assessor may develop PRGs using default values for the appropriate land uses for the site using methods described in RAGS, Part B (USEPA, 1991d). Additionally, to evaluate inter-media extrapolation, methods outlined in *Soil Screening Guidance: User's Guide* (USEPA, 1996b) and *Soil Screening Guidance: Technical Background Document* (USEPA, 1996a) may be used.

2.3.2.4 Identification of Data Needs. During the review of background information, the risk assessor will likely notice that there is limited data and information available from previous investigations, and that additional data must be collected in the RI to support a BRA. The technical team should note data gaps that exist and will need to be considered in the development of the data collection strategy for the RI. Common data gaps may include insufficient characterization of nature and extent of contamination to adequately describe an exposure pathway, insufficient background characterization, and insufficient sample number to determine a 95% Upper Confidence Limit (UCL) of the mean concentration for an exposure area.

The data needs for an RI focus on addressing the nature and extent of contamination, potential migration, and possible receptors available to complete the exposure pathways. Guided by the CSM, different types of data may be needed to address requirements and objectives of the BRA.

- Data or information in support of determining current and future land use and population characteristics.

- Data to support fate and transport modeling/calculations (total organic carbon, grain size, porosity, processed meteorological data, etc.).

- Data to conduct qualitative and/or quantitative evaluation of uncertainties in the risk assessment (mean, maximum, minimum, or the entire distribution of values for key parameters identified by a sensitivity analysis).

- Data to support qualitative assessment of potential receptors and populations (census information, postal-carrier route information/DataMap®, etc.).

- Toxicity data to assess risk or hazard. Where critical toxicity values are not available from EPA, the appropriate DOD Toxicology and Research Program offices may be consulted (e.g., USACHPPM Toxicology Directorate at: http://chppm-www.apgea.army.mil/tox/program.htm then contact the Health Effects Research Program Manager; or contact the Air Force Research Laboratory, Human Effectiveness Directorate, Operational Toxicology Division, at: http://voyager.wpafb.af.mil or (937) 255-5150 x3105).

- Representative data for evaluating the nature and extent of source and pathways, with appropriate confidence for intended data uses, and background chemical concentrations.

2.3.2.5 DQOs. The quality of a BRA is directly dependent upon the quality of the chemical data applied. Regardless of how well other components of the BRA are performed, if the quality of the data is poor or the data do not accurately reflect the site contamination or the types of exposures assessed, the BRA will not provide an adequate description of potential health effects posed by the site. Therefore, it is imperative that

the types of data scoped for use in the assessment be carefully planned.

2.3.2.5.1 Planning for appropriate data acquisition is an important step in obtaining data of the necessary quality. During this planning stage, appropriate location, numbers, and types of samples, DLs and QLs, and analytical requirements can be specified as part of the DQO process. These and other specific minimum requirements for BRA data should be specified prior to data collection by the technical team in early stages of site planning or scoping. Once available, a thorough review of the resultant data is needed to assure that the DQOs have been met (see section 4.2). This further assures that the most appropriate information is used in the BRA.

2.3.2.5.2 The risk assessor should begin to document data needed, identifying data types, location, quantity, and quality requirements. Chemical data to be collected should be identified with the appropriate QA/QC requirements. See *Guidance for Data Useability in Risk Assessment (Part A)* (USEPA, 1992h). In addition, the level of confidence (maximum error rate) required of the sample results should be set, after considering the potential variability of the sample results in a given matrix and potential laboratory/sampling handling errors. For nonchemical types of data, the QA requirements will be established on a case-by-case basis. At a minimum, the source of nonchemical data and an assessment of their reliability and representativeness for use at the site should be documented.

2.3.2.5.3 The analytical methods applied to BRA data collection should be specified as part of the minimum requirements prior to the data collection. Once data results are available, the analytical methods used and DLs and QLs attained should be reexamined to identify any deviations from the minimum requirements, and the impact of that deviation upon data useability.

2.3.2.5.4 Three broad types of analyses are available, each having a different potential use in a BRA:

- Field screening data, such as those collected with direct-reading or field instruments (photoionization detectors, combustible gas indicators, or field chemistry tests). Because of the uncertainty associated with these methods (due to lack of stringent QA/QC protocols), these data are best used qualitatively or in conjunction with verified results by more reliable methods unless the method can demonstrate equivalency with a proven method.

- Field laboratory analyses, such as those obtained from a mobile onsite laboratory. These data can be used in a BRA if appropriate QA/QC procedures have been followed and the data are of good quality, as determined by the data evaluation process.

- Definitive data. These data are appropriate for inclusion in a BRA if appropriate QA/QC procedures have been followed and the data are of good quality, as determined by the data evaluation process.

2.3.2.5.5 Several different laboratory analytical protocols are available, varying in the instrumentation, the level of QA/QC, sensitivity, QLs, and other factors. EM 200-1-3 (USACE) presents summaries of common analytical methods and identifies the instrumentation and DLs/QLs for different analytes. This resource should be consulted when choosing analytical methods to quantitate data for use in the BRA.

2.3.2.5.6 Two analytical protocols that are commonly applied to environmental sampling are the EPA's SW-846 protocol and the Contract Laboratory Program protocol. To give the USACE programs the greatest flexibility in the execution of its projects, the SW-846 methods, as published by EPA, are generally the methods employed for the analytical testing of environmental samples. These methods are flexible and can be readily adapted to individual project-specific requirements (USACE, 1994b).

2.3.2.5.7 The minimum requirements for planned BRA data collection should also specify the QLs to be attained in the chemical analyses. The limits should be low enough to enable quantitation of chemicals below concentrations of potential health concern. QLs are generally specified by the analytical method; however, deviations from planned QLs can occur as a result of matrix interferences, high chemical concentrations, laboratory variations, and other factors.

2.3.2.5.8 When selecting QLs the risk assessor and project chemist should consider that EPA risk

assessment methodology specifies that one half the sample QL should be used as the proxy chemical concentration if there is reason to believe that the chemical may be present on the site. Appropriate QLs can be determined by an evaluation of health-based screening levels for site chemicals (see paragraph 2.4.2.3).

2.3.2.5.9 Data quality. For chemical data, the level of confidence will be dependent on the experience and the ability of the laboratory to be able to deliver quality data, associated QA/QC, and variability of the chemical concentrations in the medium that was sampled. Coordination between the risk assessor and project chemist/data reviewer is recommended in order to design the sample collection program which is most likely to produce sample results with an acceptable level of confidence, considering such factors as laboratory QA/QC, level of QA required for the data, QA/QC samples, and data variability. Sensitivity requirements should be identified in this scoping phase so that the data collection program will minimize the degree of uncertainty.

2.3.2.5.10 The output of the data planning discussed above should be a SOW section and/or data needs worksheets. The purpose of documentation, as well as communication with the other team members, is to avoid potential misuse of data or the risk assessment results, making sure that the selected data collection option meets the users' and decision-makers' needs. In particular, the risk assessor should explain the minimum data quality considered to be acceptable, how nondetects are treated, and how medium-specific data are evaluated or compiled to derive/model the exposure point concentration in the risk assessment. If a health assessment, health survey, or epidemiological study is to be performed by the ATSDR, the risk assessor should (in coordination with USACHPPM for Army IRP and FUDS projects) indicate in the summary or outline how the data are to be used, evaluated, or interpreted.

2.3.2.6 Reporting Requirements. The risk assessor should define the minimum requirements associated with each of the following elements. Specification of these project study elements and minimum requirements should be recorded in the SOW. Defining minimum requirements will also add more specificity to the CSM development, allowing for easier determination of the data needs.

- Data evaluation - COPC selection, defining site-related chemicals, and nature and extent of source areas.

- Exposure assessment - pathway evaluation, fate and transport of contamination, exposure point concentration, and intake assessment.

- Toxicity assessment - determination of toxicity values.

- Risk characterization - calculation of risks.

- Uncertainty analysis - quantitative and qualitative documentation of uncertainties associated with each phase of the study.

2.3.3 Risk-Based Analysis of Remedial Alternatives; FS.

The scoping requirements for the FS focus on evaluating the potential alternatives for their effectiveness in reducing the baseline site risk. Data are needed to assess any short-term or long-term risks (if the RA lasts a duration in excess of 7 years).[18] It should be noted that many sites are required to have the RI and FS to be conducted simultaneously. Therefore the preparatory steps for conceptualizing data needs between RI and FS are comparable and will not be reiterated here.

Risk aspects of the FS are three-fold:

- Development of site-specific cleanup levels for screening remedial alternatives and consideration for adoption as RAOs.

- Evaluation of potential remedial alternatives for their abilities to meet RAOs.

- Assessment of the fate and transport mechanisms of any potential release or discharge of the media being remediated or treatment byproducts/ residues.

---

[18] The 7-year period has been suggested by EPA as the point of departure between short-term (subchronic) and long-term (chronic) risks.

In addition to evaluating the alternatives for "protectiveness" of human health and the environment, the risk-based evaluation of remedial alternatives must consider risk and toxicity reduction, interruption of the exposure pathway(s) shown to pose the principal threat in the BRA, and the post-remediation (residual) risk.

2.3.3.1 Review of Existing Data. At this project phase, the risk assessor and the project team should have a good understanding of the nature and extent of contamination. In addition, they will also have a good understanding of the site strategy and customer's goals and concept of closeout. In reviewing the background information, the risk assessor should note the AOCs requiring remediation, and the location of these areas relative to future as well as current onsite and offsite populations. Census projections and other demographic information should be reviewed. Locations of sensitive populations (nursing homes, nursery schools, etc.) should also be noted. The background information review may also identify issues of concern, for example:

- Previous or newly identified regulatory concerns relating to residual risks (risk remaining upon completion of selected remedies and/or proposed removal actions).

- Project status with respect to decision path leading to site closeout if the selected alternative is not effective or fully implemented.

- Customer's goals and objectives, plan of action, budget/time constraints for RD/RA, removal actions, and the 5-year review, if applicable.

2.3.3.2 CSM. The refined CSM developed for the BRA will be reevaluated in the FS scoping phase to account for pathways which reflect post-remediation conditions as well as pathways that may become available during remediation. Two CSMs may be developed for each remedial alternative: (1) the CSM during remediation; and (2) the CSM for the site after remediation has been completed. The former is used to guide data needs to assess short-term risks (or long-term risks if the period of remediation is in excess of 7 years); and the latter, to guide data needs for the degree of risk reduction or the post-remediation risk. The exposure pathways of concern for the short-term risk CSM are primarily air (fugitive dusts or VOC emissions) and discharge of treated effluent to ground water/surface water. It should be noted that neither of these evaluations requires an assessment of the net environmental benefit if offsite treatment/disposal is the alternative to be evaluated.[19] Therefore, the risk evaluations under FS are limited to impacts on human receptors who reside onsite or near the facility, and residual risks to receptors after implementation of the alternative. It should be noted that control measures required to mitigate short-term risks associated with remediation should be conducted in the RD/RA stage.

2.3.3.3 Identification of Data Gaps. It should be noted that this stage of HTRW project planning should focus primarily on the two questions: "What is the degree of risk reduction offered by the remedial alternative?" and "What are the potential short-term and long-term risks (if applicable) associated with implementation of the alternative?" Guided by the CSMs, data may be needed for all or any one of the following risk assessment tasks to assist in the selection of a remedial alternative:

- Data to support fate and transport modeling (e.g., grain size and processed meteorological data);

- Data to conduct qualitative and/or quantitative evaluation of uncertainties in the risk assessment (mean, maximum, minimum, or the entire distribution of values for key parameters identified by a sensitivity analysis). It should be noted that this level of effort is generally not required except for onsite incineration.

- Data to assess risk or hazard to receptors (rate, concentration, chemical identity, and toxicity) of emissions or treatment products/residues which may be released during remediation.

- Data on the treatment byproducts and residues.

2.3.3.4 DQOs. This step defines the specific data requirements according to potential exposure pathways

(ingestion of and dermal contact with ground water, inhalation of airborne contaminants, etc.) which were identified as data gaps in the previous step. SOW sections should be prepared to document required data types, locations, and quality requirements. Chemical data to be collected should be identified with the appropriate QA/QC requirements. In addition, the level of confidence (maximum error rate) of the sample results should be defined, after considering the potential variability of the sample results in a given matrix and potential laboratory/sampling handling errors. The emission or discharge data may be obtained by modeling or from the results of a performance test of the full-size model or a pilot-scale model.

2.3.3.5 Risk Calculations: RAOs and RGs. RAOs consist of medium-specific RGs, modified from PRGs during or after the BRA, to assure protectiveness of human health and the environment. The final modification to the PRGs calculates allowable media concentrations from the acceptable risk levels determined through the risk management process. RAOs should be expressed as both a contaminant level and an exposure route, as protectiveness may be achieved by either reducing the contaminant level, or by reducing or eliminating exposure. Coordination of this process with the RAOs/RGs developed during the ERA is critical to assure that the selected remedy is protective of both human and ecological receptors.

2.3.3.6 Reporting Requirements. The requirements to be reported in the FS are summarized and identified below:

- Development of RGs, presented in the RAOs section.

- Assessment of RAO protectiveness, given the acceptable risk range and uncertainties in deriving the RGs, background concentrations, and the analytical DLs. Presented as part of the screening of alternatives section.

- Assessment of long-term effectiveness/residual risk to human health and the environment (evaluate if risk reduction afforded by the proposed remedial alternatives is effective). Presented in the detailed analysis of alternatives section.

---

[19] EPA has implemented an off-site policy (USEPA, 1993a) requiring the facility receiving environmental debris or media for treatment or disposal be either in compliance with RCRA Subtitle C or under a scheduled compliance action or corrective action.

- Assessment of short-term effectiveness (evaluate if the proposed remedial options pose unacceptable short-term risks to humans onsite and offsite during the RA. Presented in the detailed analysis of alternatives section.

2.3.4 Short-Term Risks Associated With Construction. This section focuses on HTRW data scoping for the evaluation of control measures needed to mitigate short-term risks posed by construction of CERCLA removals or RAs. To meet the risk assessment or evaluation data needs, the risk assessor should coordinate with the PM, as well as other data users to identify the remedy aspects which require risk evaluation in this phase.

As a screening or comparative risk analysis has already been performed in the RI/FS project phase (or an EE/CA for a non time-critical removal action), performance of risk assessment tasks in this phase is generally limited in scope (unless there is a need for a more detailed risk assessment because the construction is likely to result in a significant release of site COCs). If this is the case, information from previously performed risk analyses should be reviewed and additional data needs identified as required. Risk assessment of removal actions or construction of the selected remedial alternative should generally follow procedures and data requirements described in RAGS Part C (USEPA, 1991e).

When evaluating data needs and their quality/quantity, consideration should be given for completing the evaluation in a timely manner. Striking a balance between the desire for site-specific/treatability data and assumed data (data from other sites) for use in the evaluation is the key aspect in this project planning stage. Other areas for project planning that may require coordination between the risk assessor and other project team members (e.g., the health and safety specialist) are:

- Short-term impact of the remedial alternatives on site environment (i.e. acute risks to ecological receptors or habitat destruction, or risks to surrounding human populations and/or on-site remedial workers).

- Risk of accidents during construction (physical hazards, explosions, spills, etc.).

- Risk communications (public perception and understanding of risks from the alternatives).

- Other risk management considerations or criteria (e.g. cost, schedule, operations and maintenance/engineering and operational flexibilities, etc.).

2.3.4.1 Review of Existing Data. The information developed in the FS in conceptualizing data needs to assess the short-term risks can be used to develop or revise the site strategy. It is recommended that the project team carefully review all site background information, RI and FS reports, and any pertinent field tests or studies.

Through qualitative or quantitative risk assessment or analyses, a determination will be made as to whether or not additional controls are needed to address risks during remediation or the residual risks. If the assessment indicates any unacceptable potential risks, the decision will focus on: (1) whether the selected remedy can be implemented under the design and operation plans without posing an unacceptable short-term risk or residual risk; (2) the need for removal actions to reduce the threat of human health risks or expedite/enhance site remediation; and (3) control measures (operational or engineering) to mitigate site risks and to assure compliance with ARARs and TBC requirements. Therefore, specific decisions associated with this executable project phase may include all or any combination of the following:

- Determine whether the selected remedial or removal actions are likely to comply with Federal and State ARARs or TBC health-based criteria required by the agencies regarding short-term risks.

- Determine if additional control measures are required to be designed and implemented to mitigate or reduce short-term risks (or if new remedies should be recommended to replace the selected remedies).

- Determine if removal actions are needed to mitigate imminent threat to human health and environment.

- Determine if the selected removal actions are consistent with the final site remedy (if such remedy is reasonably expected).

2.3.4.2 CSM. Data needed for evaluation of controls to reduce short-term risks associated with remediation should be based on the CSM developed in the FS, focusing on the potential impact of the remedy to receptors identified, and the effect of control measures. The data needed may be nonchemical in nature, e.g., engineering design parameters to reduce, remove, or change the physical/chemical nature of the emission, effluent discharge, or residues. The sources of these data may be the remediation vendors/contractor, EPA's literature (e.g., feasibility studies under the Superfund Innovative Technology Evaluation [SITE] program), or design information from other sites using the same/similar technology and wastes. The data needed may also be chemical in nature, e.g., constituent concentrations in the emissions or discharge, or the chemical identity, toxicity information, quantity, rate of release, and fate and transport characteristics of treatment byproducts, derivatives, or residues.

The CSM should be appropriately modified to help the project team focus the data collection effort to evaluate significant pathways for potential emission or discharge during the remediation period. The CSM focuses on the source, release, fate and transport, and exposure point concentration, routes and receptor to aid the risk assessment.

2.3.4.3 Identify Data Gaps. It should be noted that data needs at this stage of the HTRW project planning should primarily focus on the project decision: "What control measures are required to mitigate the short-term risk to the appropriate human receptors onsite and/or offsite (individuals and community)?" If the RA requires transportation of wastes offsite through areas of dense populations or congested transportation routes, evaluation of controls required to eliminate potential risks of accidents/spills associated with this offsite action may also be required. The risk assessor should coordinate with the health and safety specialist, design engineer, and chemist to define data quality and quantity, and locations of samples.

Guided by the CSM, data may be needed for all or any one of the following risk assessment/evaluation tasks to respond to the project decision on whether or not there is a need to impose control measures; augment or modify the selected remedy; or conduct removal actions:

- Evaluate in more detail the short-term risk assessment/analysis performed for the FS to reduce uncertainties; some of the data requirements may be:

    - Data to support fate and transport modeling/calculation, e.g., grain size of soil handled, residue or solid waste stream leaching characteristics, processed meteorological data, etc.

    - Data to assess the amount of discharge or residues, e.g, amount of soil re-suspension for a specific soil handling method, estimation of fugitive dust volatilization, stack gas emissions, or effluent discharge rates, etc. (i.e., representative monitoring or field data to assess risks and demonstrate compliance with protective criteria/standards are needed).

    - Data to support qualitative assessment of potential exposure to receptors and populations (method of residue disposal or environmental media into which effluents/emissions are discharged, transportation routes for wastes to offsite locations, population or census information, etc.).

    - Data to assess risk or hazard (toxicity information of waste residues, byproducts, derivatives, and degradation products (for bioventing or bioremediation)).

    - Data to compare ARARs and TBC short-term health goals with representative site sample or monitoring data which meet predefined QA/QC criteria.

2.3.4.4 DQOs. This step defines the data types required according to potential exposure pathways. Examples of data types according to medium for use in assessing potential exposure pathways are: incidental ingestion or dermal contact with the treatment residues or effluent and inhalation of airborne particles or volatilized organic chemicals. In each of these data types, sample data or continuous monitoring data, and data for modeling the exposure point concentration for the site contaminants or their treatment derivatives/residues in the media may be needed.

To evaluate the need for control measures for the selected remedial alternatives under this project phase for short-term impact during remediation and residual risk after remediation, data relating to the design, operation, and maintenance of the remediation system are needed to calculate the discharge or release rates of the site constituents and the process waste streams. The process waste streams include chemical characterization of all remediation or treatment byproducts, derivatives, or residues during and after remediation, which may impact onsite and offsite humans. It should be noted that the screening or comparative assessment of remedies for short-term risks should have been conducted in the FS stage, before remedy selection, and in this phase a more rigorous analysis of risks and control measures are developed for the selected alternative. The data quality used in these screening analyses should be reviewed to see if they meet the data user's requirements.

2.3.4.5 Reporting Requirements. The following presents the elements which address different aspects of controls to reduce short-term risks within the design analysis for construction of removal actions or RAs.

- The evaluation of potential control measures necessary to mitigate risks associated with remedial or removal actions; usually part of the design analysis included in the RD.

  - Health and safety design analysis; engineered barriers, monitoring, worker protection, and response measures.

  - Environmental controls and permitting; dust control, air emission control, effluent and runoff controls.

  - Methods of construction; excavation, grading, structure construction, etc. and control features associated with each.

  - Phasing of construction.

CHAPTER 3

3.0 EVALUATING THE SCREENING-LEVEL HHRA

3.1 INTRODUCTION

HHRAs performed at the PA/SI stage are typically screening-level in nature and are performed to identify whether a site needs to be assessed further or can be eliminated from further concern. Rarely would a screening-level HHRA provide adequate information to justify remediation. Since the information that is available at this point of a site response is usually limited, a conservative approach is used in performing the assessment.

3.2 SCREENING-LEVEL HHRAs

The basis of the screening-level HHRA is a comparison of site media concentrations (typically, the maximum detected concentration is used) with health-based screening levels, calculated according to RAGS protocol. The recommended values to use for performing this evaluation are those developed by EPA Region 3 (RBC Tables) or Region 9 (PRG Tables), both updated regularly. It is important to note that the RBC and PRG values noted above are not equivalent, as the exposure pathways evaluated are different. Therefore, it is imperative that these values be applied within the context that they were developed. The basis for utilizing these values will be introduced later in this chapter, and presumes an understanding of general risk assessment methodology.

PRGs are not synonymous with RGs. For a complete discussion of the development of site-specific PRGs, and appropriate methodology for calculation of RGs, see RAGS Part B (USEPA, 1991d).

3.2.1 Chemical Data Collection and Review. In order for the screening-level HHRA to achieve the desired objectives, the data applied to the assessment must be appropriate for the intended use. Data that are available from PA/SI activities are usually limited in number, but should be broad in scope of chemical analysis and in the type of media sampled.

3.2.1.1 An important component of the data review for a screening-level HHRA is an evaluation of the representativeness of the data. Sampling should have been conducted in areas of suspected contamination in order to provide information on the "worst case." If sampling was not conducted in areas of suspected contamination, the screening-level HHRA will not provide an adequately conservative assessment of potential risks. Similarly, if a broad chemical analysis was not performed, or if data are not available for all media of potential concern, the usefulness of the screening-level HHRA will be limited and would not be appropriately used to eliminate a site from further consideration.

3.2.1.2 The following factors are minimum requirements for data used in a PA/SI screening-level HHRA:

- Chemical-specific analysis of all environmental media of potential concern (e.g., soil, sediment, surface water, and ground water).

- A broad chemical analysis (or defensible historical information regarding specific COPCs).

3.2.2 Exposure Assessment. Two primary elements of the screening-level HHRA for a PA/SI are the identification of the appropriate receptor group(s) and selection of appropriate exposure point concentrations.

3.2.2.1 Selection of the population group with the highest potential exposure is required in applying the appropriate health-based screening values. Development of the preliminary CSM can be used to identify this group. The EPA regional health-based screening values are based on either residential or occupational exposures.

3.2.2.2 As a rule, the highest detected chemical concentration in a medium is compared with the health-based screening value. However, the range of chemical concentrations detected, as well as the number of samples collected, should be reviewed to determine whether this approach is appropriate. If the screening level HHRA does not provide a clear determination of whether the site can be eliminated from further consideration, further study under an RI (i.e., BRA) is indicated.

3.2.3   Health-Based Screening Levels. As noted earlier, the health-based screening levels calculated by Region 3 and Region 9 are not the same, as they evaluate different exposure pathways. The pathways evaluated are delineated as a lead in to the tables. Note that these values are updated regularly, and care should be taken to assure that the most recent values are used. The Region 3 RBC tables can be accessed on the Internet at http://www.epa.gov/reg3hwmd/risk. The Region 9 PRG tables can likewise be accessed at:

http://www.epa.gov/region09/waste/sfund/prg/index.htm

To appropriately use the health-based screening values, the risk assessor must be aware of the assumed exposure pathways and exposure factors used to derive these values. If exposure pathways other than those used for the calculations are anticipated to be significant at a given site, use of the health-based screening values is limited. Other values, developed by other EPA regions may also be appropriate, particularly if the site where the assessment is performed falls within that geographical region.

3.2.4   Risk Screening. To perform the risk screening in a PA/SI, the maximum chemical concentration in each medium is compared with the selected health-based screening level. In general, if the maximum chemical concentration exceeds the health-based screening level, further study of the site is indicated. The range of chemical concentrations detected, the degree of the exceedance of the health-based screening level, and the appropriateness of the value itself should be evaluated as part of the decision-making process in determining whether the site should be eliminated from further concern or if further study is warranted.

3.2.5   Characterization of Uncertainty. The uncertainties associated with a screening-level HHRA should be clearly presented as part of the assessment. The potential effect of the following factors should be discussed:

- Uncertainties associated with the limited chemical data base for the site.

- Use of maximum chemical concentration for representing exposure at the site.

- Use of highest exposure or "worst case" receptors.

- Application of the health-based screening value and the inherent assumptions used in its derivation.

CHAPTER 4

4.0 EVALUATING THE BASELINE HHRA

4.1 INTRODUCTION

This chapter presents the conceptual and technical objectives for evaluation of a baseline HHRA[20], and the minimum content expected to be included when evaluating a BRA. The BRA provides an objective technical evaluation of the potential health impacts posed by a site and should not incorporate policy, management, and other nontechnical factors. The BRA should be clear about the approaches, assumptions, limitations, and uncertainties inherent in the evaluation to enable the risk assessor and risk manager to interpret the results and conclusions appropriately. The BRA is used by the risk manager, in conjunction with regulatory, policy, feasibility, schedule, budget, and value of resources considerations, to determine the appropriate response actions at the site.

The BRA is one component of overall site investigative and remedial activities and, as such, should be developed with an understanding of how it is supported by preceding components of site activities, such as sampling and analysis, and how it supports and shapes follow-on components, such as remediation. Although the BRA is performed to achieve several specific objectives (such as describing potential health risks), it may also be needed to support other general response objectives.

This chapter is not intended to be a step-by-step instruction manual for developing a BRA, rather, it is a guide for reviewing and evaluating BRAs. Adequate guidance is provided in other resources for preparing a BRA, and is referred to below and throughout the chapter. This chapter discusses the important components of a BRA, highlighting where up-front planning and professional judgment are needed, and identifying the factors that should be present in a well-constructed risk assessment.

The methodology presented in this chapter has largely been developed by the EPA for activities undertaken under CERCLA. The primary guidance documents that form the basis for the discussion on BRA methodology are listed below. Of these guidance documents, RAGS (USEPA, 1989j) provides the general overview and structure of the risk assessment process. As noted earlier, a thorough understanding of RAGS is prerequisite to the USACE process, and redundancies will not be found in this guidance. This guidance will, however, provide the details necessary to focus investigations toward site closeout and provide USACE procedures relative to performance and evaluation of a site-specific BRA. Appendix A presents additional selected OSWER directives and EPA regional guidance.

- *Risk Assessment Guidance for Superfund: Human Health Evaluation Manual (Part A)* (RAGS) (USEPA, 1989j).

- RAGS Part B (USEPA, 1991d).

- RAGS Part C (USEPA, 1991e).

- RAGS Part D (USEPA, 1998a).

- *Exposure Factors Handbook* (USEPA, 1997c).

- *Guidance for Data Useability in Risk Assessment (Part A)* (USEPA, 1992h).

- *Guidance for Data Useability in Risk Assessment (Part B)* (USEPA, 1992k).

- Applicable Directives from EPA's OSWER ("OSWER Directives") (ongoing issuance), including:

  - *Guidance on Risk Characterization for Risk Managers and Risk Assessors* (USEPA, 1992d).

  - *Human Health Evaluation Manual*, Supplemental Guidance: *Standard Default Exposure Factors* (USEPA, 1991b).

---

[20] For the purposes of this text, Baseline HHRA and BRA can be used interchangeably. BRA will be used here to avoid confusion with established EPA guidance for HHRA (EPA, 1989i). It is understood that the evaluation of potential environmental risks, or ERA, is an integral part of the BRA.

- *Supplemental Guidance to RAGS: Calculating the Concentration Term* (USEPA, 1992j).

- *Revised Interim Soil Lead Guidance for CERCLA Sites and RCRA Corrective Action Facilities* (USEPA, 1994c).

• Various subject-specific guidance developed to support specific aspects of risk assessment, such as:

- *Superfund Exposure Assessment Manual* (USEPA, 1988d).

- *Dermal Exposure Assessment: Principles and Applications* (USEPA, 1992c).

- *Guidelines for Exposure Assessment* (USEPA, 1992i).

4.2 SUMMARY AND REVIEW OF ANALYTICAL DATA.

The quality of a BRA is directly dependent upon the quality of the chemical data applied. Regardless of how well other components of the BRA are performed, if the quality of the data is poor or the data do not accurately reflect the site contamination or the appropriate types of exposures, the BRA will not provide an adequate description of potential health effects posed by the site. Therefore, it is imperative that the types of data used in an assessment be carefully evaluated as well as properly used.

4.2.1   Historical Data Review. In some instances, historical data are available and can be used, in whole or in part, with or without supplemental data, to assess potential health risks associated with the site. Often, the data have been collected for purposes other than for use in a BRA and, thus, may not be appropriate for inclusion in a BRA. Prior to inclusion in a BRA, these data must be reviewed for useability.

4.2.2   Guidance. This chapter highlights several factors that should be considered when evaluating data collected specifically for a BRA, or when reviewing existing data to determine its useability. Much of the information presented herein has been obtained from the following documents:

• *Guidance for Data Useability in Risk Assessments (Parts A and B)* (USEPA, 1992h,k).

• *Laboratory Data Validation, Functional Guidelines for Evaluating Inorganics Analyses* (USEPA, 1994b).

• *Laboratory Data Validation, Functional Guidelines for Evaluating Organics Analyses* (USEPA, 1994a).

• EM 200-1-1, *Validation of Analytical Chemistry Laboratories* (USACE).

• EM 200-1-3, *Requirements for the Preparation of Sampling and Analysis Plans* (USACE).

• EM 200-1-6, *Chemical Quality Assurance for HTRW Projects* (USACE).

4.2.3   Evaluation of Data Quality. An evaluation of data quality should examine five broad categories, each discussed in the following paragraphs. The risk assessor must be aware of the important factors within each category to enable him or her to judge whether the data are appropriate for inclusion in the BRA, as specified in the DQOs. These are:

• Data collection objectives.

• Documentation.

• Analytical methods/QLs.

• Data quality indicators.

• Data review/validation.

4.2.3.1 Data Collection Objectives. The objective of the data collection program should be re-examined as part of data evaluation to determine whether the type and scope of analyses were appropriate for risk assessment purposes, and whether supportive information (such as QA/QC protocols) is available. Optimally, all data available for a BRA will have been collected with consideration of specific minimum requirements (DQOs). These data should be evaluated in terms of the attainment of these objectives or minimum requirements. Each factor specified as a minimum requirement or objective should be re-examined to determine the degree to which these requirements were attained during sampling and analysis.

4.2.3.2 Documentation. The collection and analysis of site media have been adequately documented to demonstrate that the samples were collected, handled, and analyzed according to the DQOs and/or minimum requirements specified for BRA data. Documentation on adherence to these minimum requirements should be available for review by the risk assessor.

4.2.3.3 Analytical Methods and QLs. The analytical methods, DLs, and QLs applied to BRA data collection should be specified as part of the minimum requirements prior to the data collection. Once data results are available, the analytical methods used and DLs attained should be re-examined to identify any deviations from the minimum requirements, and the impact of that deviation upon data useability.

4.2.3.4 Data Quality Indicators. Six data quality indicators (precision, accuracy, representativeness, completeness, comparability, and sensitivity) need to be considered when reviewing chemical analytical results. The assigned data evaluator/validator should examine these factors as part of the formal data evaluation procedures. However, it is important for the risk assessor to understand the terms and meaning in order to understand the data evaluation reports and how they affect the useability of the data.

4.2.3.5 Data Review/Evaluation.

4.2.3.5.1 Review and evaluation of chemical data can be performed at different levels and depths, depending on the desired use of the data. Prior to inclusion in a BRA, site data should undergo an evaluation process. Data evaluation should be performed by a chemist or other qualified individual. The risk assessor need only know that the data have been reviewed according to acceptable protocols, and all data have been appropriately qualified. Summary reports from the data evaluation will inform the risk assessor of any variations or deviations from accepted protocols.

4.2.3.5.2 Different analytical protocols have different data evaluation requirements. In addition, different protocols may use different qualifiers or

criteria for evaluating data. The risk assessor needs to be clear about the appropriate evaluation requirements for the protocols applied to assure appropriate interpretation of the data.

4.2.3.6 Data Summary/Segregation of Data. General data that have been identified as acceptable for use in a BRA should be summarized in a manner that presents the pertinent information to be applied in the BRA. Any deviations from the DQOs or minimum requirements should be identified, and the potential effects upon the BRA described in the assessment. Any data that have been rejected as a result of the data evaluation should be identified, along with a reason for their rejection. At this point in the BRA, all appropriate site data identified as acceptable by the data evaluation process should be combined for each medium for the purposes of selecting COPCs for the site, as discussed in Paragraph 4.3. However, this does not mean that all available data are to be combined. "Appropriateness" of data should take into consideration the area of exposure to be assessed.

4.3 SELECTION OF COPCs

4.3.1   Objectives. The objective of selecting COPCs for the BRA is to identify a subset of chemicals detected at the site that could pose a potential health risk to exposed receptors. The selection process is needed for several reasons:

- Not all chemicals detected at a site are necessarily related to the site. Some may be naturally occurring, a result of anthropogenic activities or of chemical use in offsite areas.

- Some chemicals may be a result of inadvertent introduction during sampling or laboratory analysis.

- Not all chemicals detected at a site are present at concentrations high enough to pose a potential exposure or health threat, or may be trace elements present at health-protective concentrations.

The chemical selection process is performed on the data that have been identified as useable by the data evaluation process. COPC selection involves evaluation of these data using a number of criteria that are designed to identify those chemicals that are not appropriate to retain as COPCs. T-hrough an exclusion process, the COPCs are selected from the list of chemicals analyzed in site media. The outcome of the selection process is a list or lists of chemicals in site media that are later assessed quantitatively in the BRA.

4.3.2   General Considerations. Two general factors should be considered before applying the chemical selection process. These factors allow the risk assessor to select the most appropriate data to include in the assessment.

• What is the exposure area?

  - Not all chemical data collected from site media represent those to which a receptor is necessarily exposed. When selecting COPCs, the potential receptors, exposure pathways, and exposure routes identified in the preliminary CSM should be examined. The preliminary CSM will identify where exposure is expected to occur (onsite, offsite, to surface soils, to subsurface soils, through ground water, by direct contact, etc.). This information is then used to help identify the media and locations where assessments will be directed and COPCs identified for each pathway of concern.

  - A distribution analysis of the chemical presence at the site should be conducted. This examination would differentiate between impacted areas and nonimpacted areas which is particularly useful at very large sites. The distributional analysis can be a statistical evaluation or performed qualitatively. The distributional analysis may identify the whole site as the exposure area or only subunits of the site as the exposure area.

• Are the chemical data appropriate?

  - Even with high quality, useable data, the form of the chemical or sampling technique should be examined for relevance for exposure. For example, unfiltered ground water data may not be relevant to exposures if all water withdrawn from an aquifer for potable purposes is normally filtered prior to consumption. Data composited from multiple locations and depths may also not be relevant to exposures if

exposure to these locations and depths is not plausible.

4.3.3   Selection Criteria/Methodology. Criteria that can be applied to determine whether a chemical should not be retained as a COPC are:

• Nondetection.

• Comparability with background concentrations.

• Non-site-relatedness.

• Role as an essential nutrient and presence at health-protective levels.

• Limited presence.

Each criterion is discussed further in the following paragraphs.

4.3.3.1 Nondetection. Chemicals analyzed for but not detected in any sample of a site medium should not be included as COPCs for that medium. Care must be taken when evaluating analytical results in which a very high DL was attained, since a significant concentration of a chemical may be "masked" due to the elevated QL. Although a quantitative estimate of the chemical's concentration value is unavailable in such a case, the chemical may be assessed qualitatively to determine if it is present in other site media (if so, EPA recommends utilizing one-half of the SQL as a proxy concentration) or re-sampling may be indicated.

4.3.3.2 Comparability with Background Concentrations.

4.3.3.2.1   Some chemicals detected in site media may be naturally occurring or present as a result of ubiquitous or offsite chemical use. Therefore, it is appropriate to exclude them from the risk assessment. Background samples are segregated from the site data, and are used exclusively to identify non-site-related chemicals.

4.3.3.2.2   Acquisition of site-specific background information is always preferable to regional or national values when examining site-relatedness and comparability to background concentrations. Literature values describing regional or national background ranges for chemicals in soil, ground water, surface water, and sediments may be used, but only if site-specific background information is unavailable. Regional or national ranges are relatively insensitive and can lead to misinterpretation of the data.

All USACE Risk Assessments Shall Include a Statistically Robust, Significant, and Defensible Set of Background Concentrations

Background values should be expressed as the 95% CL on the mean. Chemicals properly applied to the environment according to their intended use (i.e. pesticides and herbicides) shall not be considered as contaminants, but should be considered as a part of the background. In industrial areas, normal concentrations of anthropogenic contaminants shall be considered as part of the background.

4.3.3.2.3 Determination of comparability with background can be accomplished in several ways, depending on the amount of data available. Two methods that are available are statistical evaluation and numerical comparison.

- A statistical evaluation is best utilized when a sufficient number of site and background samples are available to test the null hypothesis that there is no difference between the site and background mean chemical concentrations. This approach can be used when the risk assessor has defined the minimum requirements for background and site sample numbers and sampling design. Several statistical tests are available with which to determine whether the two data groups, background and site, are comparable. Texts on statistics, such as Gilbert (1987), should be consulted for tests applicable for use in specific site conditions. The selection of test depends upon the distribution of the data (normal, non-normal), whether nondetected values are included, the number of samples, and perhaps (depending on the test) other factors. This is the most rigorous method of determining comparability.

- Numerical comparisons can be made when the background data are more limited in number,

making a statistical comparison less meaningful. This approach may be useful when historical data with limited background samples are being used, or when the minimum requirements for BRA data collection have not been met and less than optimal numbers of background sample results are available. The following comparisons can be made:

- Comparison of mean site concentration to two (USEPA, 1995d) or three (USEPA, 1992a) times the mean background concentration.

- Comparison of range of detected concentrations in both data sets.

4.3.3.5 Chemical Distribution. The physical distribution and frequency of detection of a chemical in a site medium or exposure area can be used to refine the list of COPCs. The premise behind this criterion is that a chemical with a limited presence in a medium or exposure area does not pose as great a potential health risk as do chemicals more frequently detected. The distribution of the chemical presence in a site or exposure area should be examined by identifying where the chemical was and was not detected and its frequency of detection. If this evaluation indicates that the distribution of the chemical is low, i.e., it is detected in only one or a few locations, it may be reasonable to exclude it as a COPC, or to select the chemical as a COPC for a smaller exposure area of the site. This screening should be performed in conjunction with the toxicity screening to assure that chemicals representing risks to receptors are not eliminated unnecessarily from the list of COPCs.

4.3.4 Presentation of COPCs. The conclusion of the chemical selection process is a subgroup of chemicals that are selected as COPCs and which will be used in the BRA. Tables should be developed segregating the COPCs selected for each medium and/or exposure area. All chemicals that were removed from consideration should be identified, with an explanation of the reason for their exclusion.

4.4 EXPOSURE ASSESSMENT

The purpose of the exposure assessment of a BRA is to estimate the nature, extent, and magnitude of potential exposure (or site-specific dose) of receptors to COPCs that are present at or migrating from a site, considering both current and plausible future use of the site. Several components of the exposure assessment have previously been characterized during earlier stages of the site investigation for the purposes of developing the CSM and focusing investigative activities. These components include identification of potential receptors, exposure pathways, and exposure areas. These preliminary characterizations were based upon early and often incomplete information that now must be clarified in light of the information obtained during the RI.

4.4.1 Refinement of the CSM. The CSM is a representation of certain aspects of the exposure assessment. Its earlier formulation was based upon assumptions regarding chemical presence and migration, which now should be verified and revised (if necessary) with information collected during the site investigation.

4.4.2 Characterization of the Exposure Setting.

4.4.2.1 The objective in describing the exposure setting is to identify the site physical features that may influence exposure for both current and future scenarios. While each site will differ in the factors that require consideration, some of the more common factors are listed below and discussed briefly. Examples of how the factor may influence exposures are also provided.

- Geology. The land type and forms may influence exposure in various ways. For example, the topography of the area can influence the direction of chemical migration to offsite areas. The presence of surface water bodies may indicate potential exposures through recreational or potable use of the water or through the consumption of aquatic organisms (i.e., fish and shellfish).

- Hydrogeology. The number, types, and characteristics of aquifers (depth, salinity, use, ground water flow direction, and velocity) should be examined to evaluate whether exposure to ground water is possible and, if so, where, when, and to whom.

- Climate. The temperature and precipitation profile of the area may limit the frequency of exposure (e.g., frozen surface water bodies, extent of outdoor activities) as well as influence the extent of

chemical migration (e.g., rates of volatilization and infiltration).

- Meteorology. Wind speed and direction may influence the entrainment of soil particles and the extent of transport and dilution of air contaminants.

- Vegetation. The extent of vegetation may influence the availability of soil for dermal, ingestion, or inhalation exposure and the potential for exposure through the food chain.

- Soil type. The type of soil (e.g., grain size, organic carbon, clay content) may influence soil entrainment, the degree of chemical binding, and leaching potential.

4.4.2.2 Description of the site setting in the exposure assessment should involve obtaining more specific, in-depth information than obtained during the preliminary CSM development and should be supplemented by data collected during the RI. Descriptions of portions of the exposure setting may have been discussed in other portions of the site report, and need only be referenced in this portion. However, characteristics of the exposure setting that are specific to potential exposures should be presented.

4.4.3 Identification of Exposure Pathways and Intake Routes.

4.4.3.1 An exposure pathway is the physical course a chemical takes from the source to the receptor exposed. Chemical intake is how a chemical enters a receptor after contact, e.g., by ingestion, inhalation, or dermal absorption (USEPA, 1992i). These two components are considered together in this paragraph to identify potential exposures. A complete exposure pathway consists of the following elements:

- A source and mechanism of chemical release.

- An intermedia transport mechanism (if the exposure point differs from the source).

- Migration pathway.

- A receptor group who may come into contact with site wastes.

- An exposure route through which chemical uptake by the receptor occurs.

As the field investigation has been accomplished, the chemical data can now be evaluated to determine the completeness of the pathways identified in the CSM.

4.4.3.2 Potential Exposure Routes. When performing the exposure assessment, the following exposure routes should be examined regarding the completeness of the pathway.

- Ingestion of water.

- Dermal contact with water.

- Ingestion of soil or sediments.

- Dermal contact with soil or sediments.

- Inhalation of both vapor phase chemicals and particulates.

- Exposure to biota (i.e., Ingestion of plant or animal species).

4.4.4 Identification of Potential Receptor Populations. The identification of potentially exposed receptor populations (completed during the TPP process) involves defining the current and anticipated future use of the site, and identifying the current and future activities of receptors on or near the site. At this point in the assessment, it is necessary to revisit those assumptions and evaluate whether any modifications in the preliminary assumptions are required. Chemical and physical data collected either onsite or offsite may indicate that certain receptor groups are not at risk, or that new receptors may need to be evaluated.

Future Land Uses for Risk Assessment Purposes and for Development of RAOs Shall be Land Uses that are Reasonably Expected to Occur at the Site or Facility

---

Property that is currently used for industrial or commercial purposes at facilities will most likely be used for those same purposes in the future. Even in closure situations, the land use frequently stays the same. Residential land use should not be the default land use unless it is reasonably expected to occur. It is very important the future land use be discussed early with regulators, city/county zoning officials, and the public.

4.4.5 Quantitation of Exposure (Intake or Dose). Chemical intakes, or doses, are estimated for exposures that could occur from complete exposure pathways for each receptor group. The exposures are quantified with respect to the magnitude, frequency, and duration of exposure to derive an estimate of chemical intake or site-specific dose. Intakes of chemicals are estimated by combining two general components: the chemical concentration component (or exposure point concentration) and the intake/exposure factors component. Estimation of the exposure point concentration, selection of intake and exposure factors, and specific methods of combining them mathematically are presented below.

4.4.5.1 Estimation of Exposure Point Concentrations. Exposure point concentrations represent the chemical concentrations in environmental media that the receptor will potentially contact during the exposure period. They may be derived from either data obtained from sampling or from a combination of sample data and fate and transport modeling, both of which are described below.

4.4.5.1.1 For current (and perhaps some future) exposure scenarios where the current site data are anticipated to be reasonably reflective of exposure point concentrations over the exposure period, the exposure point concentration can be directly derived from site data. For future (and perhaps some current) exposure scenarios, where current site conditions are not anticipated to be representative of exposure point concentrations over the exposure period, some form of fate and transport modeling or degradation calculations should be applied to derive these concentrations. The available data need to be examined critically to select the most appropriate data to describe potential exposure.

4.4.5.1.2   Many fate and transport models are available with which to predict exposure point concentrations from existing site data. These models are presented in other references and include the following:

- *Superfund Exposure Assessment Manual* (USEPA, 1988d).

- *Air/Superfund National Technical Guidance Study Series (Volumes I - V)* (USEPA, 1989a, 1990e, 1992o, 1993c, and 1995b).

- *A Workbook of Screening Techniques for Assessing Impacts of Toxic Air Pollutants* (USEPA, 1988h).

- *Selection Criteria for Mathematical Models Used in Exposure Assessments: Ground-water Models* (USEPA, 1988e).

- *Selection Criteria for Mathematical Models Used in Exposure Assessments: Surface Water Models* (USEPA, 1987a).

- *Rapid Assessment of Exposure to Particulate Emissions from Surface Contamination Sites* (USEPA, 1985).

- *Methodology for Assessing Health Risks Associated with Indirect Exposure to Combustor Emissions* (USEPA, 1990b).

4.4.5.1.3   The type of model and level of effort expended in estimating exposure point concentrations with a model should be commensurate with the type, amount, and quality of data available. In general, it is best to begin with a model that employs simplified assumptions (i.e., a "screening level" approach) and determine whether unacceptable health risks are posed by the exposure point concentration estimated by this approach. If so, a more complex model that applies less conservative assumptions should be used to then derive the exposure point concentrations.

A Minimum of Two Risk Estimates Should be Presented for Each Land Use Scenario: the RME and the CT.

The goal of the BRA is to provide information on potential risks presented by contamination for risk managers to make informed decisions regarding future action. The risk manager needs more information than just worst case to make a good risk management decision. Multiple exposure scenarios within a land use paradigm should be used in the risk assessment to provide the risk manager with information relative to ranges of the perceived risks.

In order to describe a range of potential exposures presented by a site, the BRA should assess more than one potential exposure scenario. Use of a single expression of potential health risks does not provide information on the possible range of health risks, and does not allow the risk manager to evaluate the "reasonableness" of the estimate. Current risk assessment guidance suggests assessing an exposure scenario that represents the high end of the risk distribution, relating to a 90th percentile exposure (often referred to as an RME scenario), and a scenario which more closely describes an average exposure (or CT) (USEPA, 1-992d). Presentation of both (and perhaps additional) scenarios provides information about the range of potential risks.

4.4.5.1.4   Numerous sources are available to select appropriate intake and exposure factors for use in a BRA (see Section 4.1 for the primary EPA guidance documents). In addition to these general references, some EPA regional offices and state environmental or health agencies have developed exposure risk assessment guidance to supplement the EPA Federal guidance.

4.4.5.1.5   Some of the EPA documents provide ranges of values for intake and exposure factors, while others present values intended to represent a specific exposure. For example, the *Standard Default Exposure Factors* (USEPA, 1991b) was developed as guidance only, and the values are intended to be used when site-specific information is not available. EPA encourages the use of site-specific data so that risks can be evaluated to more closely reflect site-specific exposures. Default values should be used to calculate a high end exposure only when there is a lack of site-specific data or alternate values cannot be justifiably supported.

The Exposure Assessment of a BRA Shall Utilize Site-Specific Frequencies and Durations Whenever Possible.

Where possible, the BRA should use site-specific parameters for input into the risk algorithms. By the use of these parameters, the BRA will tailored to the actual expected exposures. Additionally, anticipated ranges of values may be used when the BRA utilizes probabilistic methods.

4.4.5.1.6   All values that are used in estimating chemical intake should be clearly presented in the assessment, the source of the value should be identified, and the rationale for using the value provided.

4.4.5.2   Calculation Methodology.

4.4.5.2.1   RAGS identifies general intake equations for each exposure pathway and should be consulted when performing the intake assessment. Some overall assumptions in the use of these equations are presented in the following paragraphs.

4.4.5.2.2   The intake equations developed by EPA for the ingestion and inhalation pathways do not contain a factor to account for bioavailability and, therefore, may predict an intake higher than one that would occur in actual circumstances. By not including a factor to consider bioavailability, it is assumed that 100 percent of the chemical detected in the medium is bioavailable. Modifications may sometimes be made to these intake equations to account for this factor, if the appropriate information is available.

4.4.5.2.3   Bioavailability refers to the ability of a chemical to be "available" in the body to interact and have an effect. There are many aspects to bioavailability; however, the type most of concern to BRAs is the ability of the chemical to be absorbed into the body. Although the medium in which the chemical is contained may be contacted, the chemical may not be

absorbed for a number of reasons, including the chemical form, competition with other factors (e.g., food in the stomach), damage of the organ (e.g., stomach, lung), effect of the medium in which the chemical is contained, and others. Many of these cannot be reliably addressed in a BRA; however, two of these can, the chemical form and the effect of the medium on absorption.

4.4.5.2.4   The form of the chemical can affect the degree of absorption into the body. This factor is most important for chemicals that form compounds (such as metals and cyanide) and chemicals that exist in different valence states (again, some metals). For example, soluble compounds of metals such as barium sulfate are readily absorbed through the stomach, whereas insoluble forms such as barium carbonate are usually not absorbed. Usually, when environmental media are analyzed, chemicals are reported as an isolated entity (e.g., barium), and no information is provided on valence state or compounds that existed in the medium. However, if the form of the chemical used at the site is known, and information on the absorption of that chemical form is available, the intake equation can be modified to account for a specific absorption.

4.4.5.2.5   The medium in which the chemical is contained also can affect the degree of bioavailability. This is most pronounced in media that demonstrate an ability to bind chemicals (such as soil and sediments). When ingested or inhaled into the body, a competition occurs between retention of the chemical on the medium and absorption of the chemical into the body. Therefore, some of the chemical may be excreted from the body without having been absorbed and some may have been absorbed and available to exert an effect. Many factors can influence the degree to which the medium will bind the chemical, most of which cannot be reliably predicted (for example, nature of the medium [organic carbon or clay content, particle size], other chemicals being absorbed, pH, organ condition, etc.). In some instances, information may be available on the degree to which a particular medium affects specific absorption routes, and the equation can be modified to account for these influences.

4.4.5.2.6   In most assessments, it is assumed that the chemical concentrations remain constant over time, often for as long as 30 years. In many cases, this assumption will not be valid. Chemical concentrations are usually reduced over time by degradation, migration, dilution, volatilization, or other removal processes. If the appropriate site-specific characteristics for natural attenuation (e.g., soil properties, climate, pH, grain size, etc.) are known and can be quantified, a concentration that decreases over time can be derived for assessing intakes through modeling.

4.4.5.3 Assessment of Uncertainties. At the conclusion of the exposure assessment, the uncertainties associated with the estimation of chemical intake should be summarized. The basis for the uncertainty should be identified (e.g., use of a default parameter), the degree of the uncertainty qualitatively estimated (low, medium or high), and the impact of the uncertainty stated (overestimate and/or underestimate).

4.5 TOXICITY ASSESSMENT

4.5.1 Objectives. The toxicity assessment fulfills two objectives in a risk assessment. First, it results in the selection of appropriate toxicity values to use in generating estimates of potential health risks associated with chemical exposure. This is accomplished by identifying appropriate sources of toxicity values and reviewing the available information to identify the most appropriate values to use. Second, the toxicity assessment forms the basis for developing summaries of the potential toxicity of the COPCs for inclusion in the risk assessment. This is accomplished by reviewing the available information on the toxicity of the COPCs and summarizing the factors pertinent to the exposures being assessed.

4.5.2 Derivation of Toxicity Values. Most toxicity values applied to risk assessments have been developed by EPA and generally do not need to be developed by the risk assessor. However, to appropriately select and use toxicity values, and to identify assumptions and uncertainties associated with them, an understanding of the development is needed. For a complete discussion of this procedure, see RAGS (USEPA, 1989j).

4.5.3 Toxicity Assessment for Carcinogenic Effects.

4.5.3.1 The toxicity value used to describe a chemical's carcinogenicity is the cancer slope factor (SF). Two types of SFs are available: oral SFs and inhalation SFs,

and are expressed in terms of $(mg/kg-dy)^{-1}$. EPA's Human Health Assessment Group reviews the SFs developed by different EPA program offices to reach an agency consensus on the value and to verify the SF.

4.5.3.2 In addition to the numerical value, each potentially carcinogenic chemical is assigned a "weight of evidence" category, expressing the likelihood that the chemical is a human carcinogen. Six categories exist (A, B1, B2, C, D, and E). In general, carcinogenic assessments are performed for chemicals in groups A, B1, B2, and on a case-by-case basis in group C.

4.5.4 Toxicity Assessment For Noncarcinogenic Effects.

4.5.4.1 Chemicals that cause toxic effects other than cancer such as organ damage, physiological alterations, and reproductive effects are generically grouped as noncarcinogens. These types of toxicants share one point in common in regard to their effects: the apparent occurrence of a toxicological threshold. This threshold is an exposure level that must be exceeded for the adverse impact of the chemical to manifest itself. Below this threshold, factors such as the body's protective mechanisms (e.g., metabolism, elimination) can limit the chemical effects, preventing the expression of adverse effects. The basis of the derivation of noncarcinogenic toxicity values, then, is to identify this threshold level, and modify it to express potential human toxicity.

4.5.4.2 The toxicity descriptor most commonly used in risk assessments for describing a chemical's noncarcinogenic toxicity is the reference dose (RfD) or reference concentration (RfC). An RfD or RfC "is a provisional estimate (with uncertainty spanning perhaps several orders of magnitude) of a daily exposure to the human population (including sensitive subgroups) that is likely to be without an appreciable risk of deleterious effects during a portion of a lifetime, in the case of a subchronic RfD or RfC, or during a lifetime, in the case of an RfD or RfC" (USEPA, 1992e).

4.5.4.3 Several types of RfDs are available:

- Chronic RfDs, used to assess chronic exposures (greater than 7 years [one-tenth of a lifetime]). Two different types of chronic RfDs are available: oral

RfD$_o$ and inhalation RfD$_i$. More recently, RfCs have been developed for the inhalation route.

- Subchronic RfD$_s$, for exposures between 2 weeks and 7 years. Both oral and inhalation subchronic RfDs (RfD$_{so}$ and RfD$_{si}$, respectively) may be available.

- Developmental RfD$_{dt}$, used to evaluate potential effects on a developing organism following a single exposure event (very few have been developed).

4.5.4.4 EPA's RfD workgroup reviews and verifies existing chronic RfDs and develops new RfDs, and resolves conflicting toxicity values developed within the EPA in the past. The RfD workgroup also states the degree of confidence associated with the study, the database, and the RfD (low, medium, or high). Subchronic RfDs are not reviewed or verified and are, therefore, considered unverified values. These values should only be used when chronic RfDs are not available.

4.5.5   Sources of Toxicity Values.

4.5.5.1   Several sources of up-to-date toxicity values and supplementary information are available. These sources are presented below. A hierarchical approach is recommended when consulting these sources: if information is not available through the first source, the second should be consulted, and so forth.

- Integrated Risk Information System (IRIS). This is EPA's primary database for the reporting of up-to-date toxicity values that have been verified by the EPA. IRIS may be accessed through the Internet at http://www.epa.gov/ngispgm3/iris/index.html. IRIS contains chemical profiles that present verified chronic RfDs, chronic RfCs, and cancer SFs. The study(s) from which the toxicity value was derived is summarized, and the method of derivation is explained (e.g., applied uncertainty and modifying factors, level of confidence, extrapolation model). Supplementary toxicity information is also sometimes included. In addition, some IRIS files contain regulatory information (such as the SDWA Maximum Contaminant Levels [MCLs] and CWA Ambient Water Quality Criteria), and often

chemical and physical properties, synonyms, and other information.

- Health Effects Assessment Summary Tables (HEAST). This document is published annually by EPA and is a collection of interim and provisional toxicity values developed by EPA. Verified toxicity values are not presented in the most current version of HEAST, rather, the user is directed to IRIS. HEAST can be obtained through the National Technical Information Service (NTIS).

- EPA's Superfund Health Risk Technical Support Center (513-569-7300). Assistance may be requested from these offices on the existence of provisional toxicity values not presented in either IRIS or HEAST or on other factors relating to risk assessment. However, EPA only provides services for sites being managed under the Federal Superfund Program.

- For sites other than Superfund, the USACE user is directed to contact the appropriate DOD Toxicology and Research Program offices: USACHPPM Toxicology Directorate at: http://chppm-www.apgea.army.mil/tox/program.htm, then contact the Health Effects Research Program Manager; or contact the Air Force Research Laboratory, Human Effectiveness Directorate, Operational Toxicology Division at: http://voyager.wpafb.af.mil or (937) 255-5150 x3105.

4.5.5.2   Additional information on the toxicity of the chemicals can be found in the following general sources:

- EPA criteria documents such as those regarding drinking water, ambient water, and air quality, as well as health effects assessment documents.

- Toxicological Profiles developed by ATSDR.

4.5.6   Use of Toxicity Values. Toxicity values developed by EPA can generally be used directly in a risk assessment with few or no modifications. The mechanism for combining toxicity values with exposure or intake estimates is described in Section 4.7. However, there are a number of factors that should be considered when applying these toxicity values. These are discussed in the following paragraphs.

4.5.6.1 Absorption Considerations. Most toxicity values are based on administered, rather than absorbed, doses, and the absorption efficiency has not been considered. However, whatever absorption has occurred during the toxicological study is usually inherent in the toxicity value. Therefore, use of a toxicity value assumes that the extent of absorption observed in the study is also appropriate for the exposure pathway being assessed. Differences in absorption efficiencies between that applicable to the toxicity value and that being assessed may occur for a number of reasons. Two factors that will influence absorption efficiencies are differences in chemical form and differences in the exposure medium.

4.5.6.2 Use of Oral Toxicity Values for Assessment of Dermal Exposure Route. EPA does not generate toxicity values for dermal exposures. As a surrogate, oral toxicity values are applied to the assessment of dermal exposures. However, since dermal intakes are based upon absorbed doses and most oral toxicity values are based upon administered doses, the oral toxicity value may be modified before using in a dermal assessment. For a complete discussion of this procedure, when it should be used, and the appropriate procedures for its application, see Appendix A of RAGS (USEPA, 1989j).

4.5.7 Special Chemicals. Some chemicals commonly detected at a site require a specific methodology to generate a toxicity value or are reported in a manner that influences the toxicity value. The following chemicals are discussed relative to these special circumstances:

- Lead.

- PAHs.

- Polychlorinated biphenyls (PCBs).

- Chlorinated dibenzo-p-dioxins and dibenzofurans (CDDs/CDFs).

- Total petroleum hydrocarbons (TPH) and other petroleum groupings.

- Military unique chemicals.

4.5.7.1 Toxicity Values for Lead.

4.5.7.1.1 Lead is a unique chemical in its pharmacokinetic and toxicological properties. Although classified as both a potential carcinogen (B2 weight of evidence) and a noncarcinogen, lead is most often assessed as a noncarcinogen only, since these effects manifest themselves at doses lower than those for carcinogenicity. However, in contrast to the assumption of the existence of a threshold for noncarcinogenic responses, there does not appear to be a threshold below which lead does not elicit a response. For these reasons and others (including lead's propensity to accumulate in bone tissue), the use of blood lead (PbB) levels, rather than chronic daily intakes, is the best indicator of potential adverse impacts). EPA has not developed a noncarcinogenic RfD or a carcinogenic SF for lead.

4.5.7.1.2 EPA has developed an exposure model for lead that considers both its biokinetics and toxicological properties. The IEUBK model (Pub. #9285.7-15-2, PB93-963511) is available through NTIS. The model integrates the intake of lead from multiple sources, including soil, food, and water ingestion, inhalation, and, when appropriate, maternal contributions. Intakes are assessed for children from the ages 0 (birth) to 7. The model does not assess lead intakes for older children or adults. Childhood exposure to lead is the focus of this assessment because this receptor group is recognized as the most sensitive to the noncarcinogenic effects of lead.

> **Use of the EPA'S IEUBK Model for Lead Exposures Should be Limited to Residential, Childhood Exposures Only.**
>
> ───────────────────
>
> Where adult and/or non-residential exposures are expected, a more appropriate model should be used. See *Recommendations of the Technical Review Workgroup for Lead for an Interim Approach to Assessing Risks Associated with Adult Exposures to Lead in Soil* (USEPA, 1996c).

4.5.7.1.3 The IEUBK model integrates intakes of lead from multiple exposure routes and predicts a PbB level, in μg/dL, at different ages (up to 7 years of age). The maximum predicted PbB level can then be compared with a threshold level of 10 μg/dL, which EPA has adopted as an "acceptable" PbB level.

4.5.7.1.4   Use of the IEUBK model is recommended when children of this age group are anticipated to be receptors at a site. However, when adults are the only potential receptors, the EPA's Technical Review Workgroup for Lead has developed an interim approach for evaluating adult soil lead exposure. *Recommendations of the Technical Review Workgroup for Lead for an Interim Approach to Assessing Risks Associated with Adult Exposures to Lead in Soil* (USEPA, 1996c) provides the currently accepted methodology. This interim guidance is available on the Internet at: http://www.epa.gov/superfund/oerr/ini_pro/lead.

4.5.7.2   Toxicity Values for PAHs.

4.5.7.2.1   PAHs, also known as polynuclear aromatic hydrocarbons or polynuclear aromatics, are a class of compounds containing hydrogen and carbon in multiple ring structures. There are numerous possible PAH molecules, many of which are commonly analyzed for in a semivolatile chemical analysis.

4.5.7.2.2   PAHs are a natural component of petroleum and are found in heavier petroleum fractions such as lube oil, naphtha, jet fuel, etc. PAHs are also produced by the incomplete combustion of organic matter, and are created during fires, volcanoes, combustion of gasoline, burning of wood, etc. For these reasons, PAHs are ubiquitous in the environment at low levels, particularly in soil and sediments, to which they readily bind.

4.5.7.2.3   Some PAHs are classified by EPA as potential human carcinogens, including:

- Benzo(a)anthracene.

- Benzo(a)pyrene.

- Benzo(b)fluoranthene.

- Benzo(k)fluoranthene.

- Chrysene.

- Dibenzo(a,h)anthracene.

- Indeno(1,2,3-cd)pyrene.

4.5.7.2.4   EPA has developed a cancer SF for one carcinogenic PAH only: benzo(a)pyrene. However, comparative toxicity values have been proposed for the other carcinogenic PAHs that describe the toxicity relative to the toxicity of benzo(a)pyrene. Several sets of comparative toxicity values have been proposed. The EPA's *Provisional Guidance for Quantitative Risk Assessment of Polycyclic Aromatic Hydrocarbons* (USEPA, 1993b) should be consulted for Toxicity Equivalence Factors (TEFs) to utilize in this assessment.

4.5.7.2.5   Other PAHs are considered by EPA to be noncarcinogens; however, only a few of these currently have RfDs. Currently, there is no comparative toxicity approach for estimating the toxicity of noncarcinogenic PAHs that do not have RfDs.

4.5.7.3   Toxicity Values for PCBs.

4.5.7.3.1   PCBs are a group of chlorinated compounds based on the biphenyl molecule. There are 209 possible individual congeners of PCBs, differing in the degree and location of chlorination. PCBs are seldom analyzed as individual compounds; rather, they are commonly analyzed as total PCBs, Aroclor compounds (a commercial mixture, with Aroclor™ being Monsanto's trade name) or sometimes in congener groups (such as tetrachlorobiphenyls or pentachlorobiphenyls). When analyzed as Aroclors, the results are expressed relative to different commercial mixtures of Aroclor, such as Aroclor 1248, Aroclor 1254, or Aroclor 1260.

4.5.7.3.2   The toxicity values (cancer SF and RfD) developed for PCBs are based on specific Aroclor mixtures -- the SF is based on Aroclor 1260 and the RfD of Aroclor 1016. These values are used to assess the potential impacts of PCBs reported in any form (i.e., another Aroclor mixture or total PCBs). However, it is known that the toxicity of PCBs varies between these congeners. Most notably, the carcinogenic potency is less in smaller molecular weight chlorinated biphenyls. Therefore, application of the Aroclor 1260 cancer SF to

Aroclor 1232 or 1248 mixtures may overestimate the degree of health risk posed by the PCB.

4.5.7.3.3   EPA recommends the use of a tiered approach to the evaluation of PCB carcinogenicity, even though toxicity values for the different Aroclors are still available. Information on the application of this procedure can be found on the IRIS database, accessible on the Internet at:
http://www.epa.gov/ngispgm3/iris/index.html.

4.5.7.4   Toxicity Values for CDDs/CDFs.

4.5.7.4.1   CDDs/CDFs, often abbreviated "dioxins and furans," are a group of chlorinated compounds based on the dibenzo-p-dioxin or dibenzofuran molecule (both of which are structurally similar). CDDs/CDFs are not compounds used for commercial purposes in the past, and, outside of research, have no known use. Rather, CDDs/CDFs are byproducts of high temperature combustion of chlorinated compounds and impurities in other chemical products such as pentachlorophenol or PCBs. Although not considered a "natural" product, some forms of CDDs and CDFs (specifically octa-CDD and octa-CDF) are ubiquitous in the environment at very low concentrations.

4.5.7.4.2   There are 75 possible CDD congeners and 135 possible CDF congeners. As with PCBs, the degree of toxicity varies with the degree and location of the chlorine atoms on the hydrocarbon ring, becoming higher when the 2, 3, 7, and 8 positions of the molecule have chlorine atoms. Considered the most potent CDD, 2,3,7,8-tetrachlorodibenzo-p-dioxin (2,3,7,8-TCDD) is the reference against which all other CDDs and CDFs are compared.

4.5.7.4.3   Analysis of CDDs and CDFs is most commonly reported by congener group (i.e., as either tri-, tetra-, penta-, hexa-, hepta-, or octachlorodibenzo-p-dioxin or -dibenzofuran). Within these groups the results are often further separated into "2,3,7,8- substituted" or "other" categories. This form of reporting is needed to appropriately assess CDDs and CDFs. Reporting as "total dioxins" or even just by congener group may require the assumption that all CDDs/CDFs present are as toxic as 2,3,7,8-TCDD, resulting in an overestimate of potential health risks posed by the presence of CDDs/CDFs.

4.5.7.4.4   A toxicity value (cancer SF) is available for 2,3,7,8-TCDD. As a policy, EPA has developed a TEF approach for other CDDs/CDFs, wherein the toxicities of these other compounds are expressed relative to the toxicity of 2,3,7,8-TCDD. These values can be used to express the amount of CDDs/CDFs present in a sample as "2,3,7,8-TCDD equivalents." Further discussion of the TEFs for CDDs/CDFs can be found in USEPA, 1989d.

4.5.7.5   Toxicity Values for TPHs and Other Petroleum Groupings.

4.5.7.5.1   Use of chemical-specific data to derive an estimate of potential health risks is the recommended method of performing a BRA. Use of chemical groupings such as TPH is less than optimal, since these types of chemical groupings vary in their chemical composition and, hence, toxicity.

4.5.7.5.2   Some attempts have been made to derive toxicity values for TPH. However, since the composition of TPH varies from place to place (even within the same site) with the age of the spill, and the type of fuel spilled or disposed, it is unlikely that these estimates are valuable descriptors of the potential toxicity of the components comprising the TPH detection.

4.5.7.5.3   For some other chemical groupings, toxicity tests have been performed on the specific mixture, and adequately describe the toxicity of the chemical grouping, such as jet fuel and diesel fuel. One potential pitfall to using these values is that the RfD may represent the toxicity of the mixture when fresh, but may not represent the toxicity of the mixture after release to the environment. When released, processes such as biodegradation, chemical migration, and transport may alter the composition of the mixture, making it more concentrated in some compounds and less concentrated in others. In these instances as well, chemical-specific analysis of the media is preferred.

4.5.7.6   Toxicity values for Military Unique Chemicals. Many DOD sites contain potentially toxic chemicals not commonly found except on military sites. Military unique chemicals may include explosives, rocket fuels, radioactive materials, chemical agents, or degradation products of these compounds. Because of the unique status of many military compounds, EPA is often unable to supply toxicity information. Toxicity information can usually be obtained by contacting the USACHPPM Toxicology Directorate at: http://chppm-www.apgea.army.mil/tox/program.htm, then contact the Health Effects Research Program Manager.

4.6 RISK CHARACTERIZATION

4.6.1   Objective. In the risk characterization, the chemical intakes estimated in the exposure assessment are combined with the appropriate critical toxicity values identified in the toxicity assessment. The results are the estimated cancer risks and noncarcinogenic health hazards posed by the exposures. Along with the numerical estimates of potential health risks and hazards, a narrative describing the primary contributors to health risks and hazards and factors qualifying the results are presented.

4.6.2   Methodology. In the following paragraphs, the methodology is presented for performing the quantitative risk characterization for carcinogens, followed by the methodology for noncarcinogens. These are discussed separately because different methodologies are used for each of these classes of chemicals.

4.6.2.1 Carcinogenic Risks. The objective of a risk characterization for carcinogenic chemicals is to derive an estimate of the overall cancer risk associated with exposure to all potential carcinogens at a site through all routes of exposure for a given receptor group, for both CT and RME current and future use scenarios. To derive this value, the cancer risk associated with exposure to a single carcinogen through a single exposure pathway is estimated. These single chemical risk estimates are then combined (added) within a pathway to describe the risk associated with a given pathway. Pathway-specific risks are then combined (added) for all exposure pathways for a given receptor group to derive an overall risk estimate for each of the cases.

4.6.2.2 Noncarcinogenic Hazards.

4.6.2.2.1   The objective of a risk characterization for noncarcinogenic chemicals is to compare the estimated chemical intake of one chemical through one exposure route with the "threshold" concentration; that is, the

level of intake that is recognized as unlikely to result in adverse noncarcinogenic health effects (i.e., the RfD). The comparison of estimated intake and acceptable exposure level is called a hazard quotient (HQ).

4.6.2.2.2   An HQ of 1 indicates that the estimated intake is the same as the RfD, whereas an HQ greater than 1 indicates the estimated intake exceeds the RfD. No further conclusions can be drawn as the relationship between intake and toxicity used to derive the RfD is not linear. In contrast to cancer risk estimates, HQs can range from values less than 1 to greater than 1.

4.6.2.2.3   To examine the potential for the occurrence of adverse noncarcinogenic health effects as a result of exposure to multiple noncarcinogens through multiple exposure pathways (for each of the exposure scenarios; current-future for average and upper bound exposures), it is assumed that an adverse health effect could occur if the sum of the HQs exceeds 1. In other words, even if exposure to each individual chemical is below its RfD (HQ less than 1), if the sum of the ratios for multiple chemicals exceeds unity, adverse health effects could occur.

4.6.2.2.4   Applying the assumption of additivity is considered to be a conservative approach, but may overestimate or underestimate the actual potential health risk presented by the exposure. If the overall hazard index (HI) is greater than unity, consideration should be given to the known types of noncarcinogenic health effects posed by exposure to the chemicals. If the assumption of additivity is not valid (i.e., if the chemicals most strongly contributing to the exceedance of the HI display very different types of noncarcinogenic effects) the HI may be segregated according to toxicological endpoint. These segregated HIs may then be examined independently.

4.6.2.2.5   Factors that need to be considered in segregation of endpoints include the critical toxicological effect upon which the toxicity value is based, as well as other toxicological effects posed by the chemical at doses higher than the critical effect. Major categories of toxic effects include neurotoxicity, developmental toxicity, immunotoxicity, reproductive toxicity, and individual target organ effects (hepatic, renal, respiratory, cardiovascular, gastrointestinal, hematological, musculoskeletal, dermal, and ocular) (USEPA, 1989j).

4.7 EVALUATION OF UNCERTAINTIES AND LIMITATIONS

4.7.1   Objective.

4.7.1.1   EPA has identified two requirements for full characterization of risk. First, the characterization must address qualitative and quantitative features of the assessment. Second, it must identify any important uncertainties in the assessment. Methods of identifying and describing uncertainties in a risk assessment are discussed below.

4.7.1.2   According to recent guidance (USEPA, 1992d):

"EPA risk assessors and managers need to be completely candid about confidence and uncertainties in describing risks and in explaining regulatory decisions. Specifically, the Agency's risk assessment guidelines call for full and open discussion of uncertainties in the body of each EPA risk assessment, including prominent display of critical uncertainties in the risk characterization. Numerical risk estimates should always be accompanied by descriptive information carefully selected to assure an objective and balanced characterization of risk in risk assessment reports and regulatory documents."

4.7.1.3   Identification and discussion of uncertainty in an assessment is important for several reasons (USEPA, 1991a):

- Information from different sources carries different kinds of uncertainty, and knowledge of these differences is important when uncertainties are combined for characterizing risk.

- Decisions must be made on expending resources to acquire additional information to reduce uncertainties.

- A clear and explicit statement of the implications and limitations of a risk assessment requires a clear and explicit statement of related uncertainties.

- Uncertainty analysis gives the decision-maker a better understanding of the implications and limitations of the assessments

4.7.2   Sources of Uncertainty.

Sources of uncertainty exist in almost every component of the risk assessment. Overall, uncertainties can arise from two main sources: variability and data gaps. Uncertainty from variability can enter a risk assessment through random or systematic error in measurements and inherent variability in the extent of exposure of receptors. Uncertainty from data gaps is most prominently seen when approximations are made regarding exposures, chemical fate and transport, intakes, and toxicity. Specific sources of uncertainty in a risk assessment are identified and discussed below. Following this discussion, different approaches for conducting an uncertainty evaluation are presented.

4.7.2.1   Uncertainties Associated with Sampling and Analysis.

4.7.2.1.1   The identification of the types and numbers of environmental samples, sampling procedures, and sample analysis all contain components that contribute to uncertainties in the risk assessment. Decisions regarding the scope of sampling and analysis are often made based on the CSM developed at the planning stages of the investigation. While appropriate planning may minimize the uncertainty associated with these components, some uncertainty will always exist, and cannot always be reduced realistically, rather it may be sufficient to just understand the degree of uncertainty associated with the assessment.

4.7.2.1.2   Some of the assumptions in this component that contribute to uncertainty in the assessment include:

- Media sampled. Due to budget limitations, only representative areas of the site are selected for sampling and analysis. This selection is usually based upon the anticipated presence of a chemical in a medium from the site history and the chemical's chemical and physical properties. If all areas of the site in which a chemical is present have not been sampled, small incremental risks either less than or equal to the risk accounted for in the BRA may not be described, although this approach is usually not feasible.

- Locations sampled. The type of sampling strategy selected may impact the uncertainty associated with the results. For example, purposive sampling (sampling at locations assumed to contain the chemicals) will likely result in a higher frequency of chemical detection and concentration than random sampling or systemized grid sampling. Therefore, use of the results may skew the assessment toward greater assumed exposures.

- Number of samples. Fewer samples result in a higher degree of uncertainty in the results. This is demonstrated in the summary statistics, specifically the 95% UCL, in which the statistical descriptor ("t" or "H" value), and hence the 95% UCL, increases with a lesser number of samples. Planning for a specific number of samples to reach a specific degree of statistical confidence can limit the degree of uncertainty, although reduction may not be feasible and quantifying the uncertainty may be just as effective in defining risks.

- Sampling process. The sampling process itself can contribute to uncertainties in the data from a number of factors, including sampling contamination (cross-contamination from other sample locations, introduction of chemicals used in the field), poorly conducted field procedures (poor filtering, incomplete compositing), inappropriate sample storage (headspace left in containers of volatile sample containers, inappropriate storage temperatures), sample loss or breakage, and other factors. Some of these factors can be controlled by an adequate SAP; however, planning does not prevent the occurrence of sampling errors.

- Analytical methodology. The analytical methodology can contribute to uncertainty in a number of ways, including the chemicals analyzed (if analyses of all important chemicals were not performed), the DLs or QLs applied (if not sufficient), limitations in the analysis due to matrix effects, chemical interferences, poorly conducted analyses, and instrumentation problems. Some of these factors can be addressed in up-front planning (such as selection of the analytical method), others cannot (instrumentation problems) be mitigated.

4.7.2.2 Uncertainties Associated with Selection of COPCs. Evaluation of the data to select COPCs for the risk assessment may result in uncertainties. Application of selection criteria may inadvertently result in an inappropriate exclusion or inclusion of chemicals as COPCs. Improper inclusion or exclusion of chemicals can result in an underestimation (if inappropriately removed) or overestimation (if inappropriately retained) of potential health risks. Uncertainties associated with the selection criteria include the following:

- Background comparison. If background measurements are not truly representative of background conditions, chemicals may be inappropriately retained or removed from the list of COPCs.

- Sample contamination. Uncertainty in the assessment can occur if chemicals are not recognized as being present as a result of sampling or laboratory introduction and are included as COPCs.

- Frequency of detection. Use of detection frequency as a selection criterion may result in the inappropriate exclusion of chemicals as COPCs.

4.7.2.3 Uncertainties Associated with the Exposure Assessment. Exposure estimates are associated with a number of uncertainties that relate to the inherent variability of the values for a given parameter (such as body weight) and to uncertainty concerning the representativeness of the assumptions and methods used.

- Potential exposure pathways. Potential exposure pathways are identified by examining the current and future land uses of the site and the fate and transport potential of the COPCs. While current land use and potential exposure pathways are often easy to identify, potential future uses can only be inferred from information available. For these reasons, sometimes the most conservative potential future land use (i.e., residential) is often assumed in many assessments to avoid underestimating potential health risks. This and any assumption regarding future land use and exposure pathways will add uncertainty to the assessment.

- Potentially exposed receptors. As discussed above, identification of potentially exposed receptors is based upon information currently available. Assumed exposed receptors under future use scenarios can only be obtained from census projections, land planning, and ownership records and can add uncertainty to the assessment.

- Exposure and intake factors. Point values for exposure estimates are commonly used in risk assessments rather than a distribution of exposure values that describe the distribution of exposures. These values are usually conservative, and their use results in introduction of conservatism into the risk assessment. Conversely, use of average (CT) and the upper end (RME) exposure and intake factors describing a range of exposures may reduce this conservativeness. Additionally, selection of site-specific exposure and intake factors will lessen the uncertainty to some degree, but since not all potentially exposed receptors will be exposed to the same degree, uncertainty cannot be eliminated.

- Exposure point concentrations. Exposure point concentrations are derived from measured site media chemical concentrations alone and fate and transport modeling. With regard to estimating exposure point concentrations from sampling data alone, use of 95% UCL and mean concentrations is associated with some degree of uncertainty. The 95% UCL is used to limit the uncertainty of estimating the true mean concentration from the sample mean concentration. This value may overestimate the true mean concentration. Use of the sample mean concentration may under- or overestimate the true mean concentration. Therefore it is strongly recommended that both values are used to represent a range of exposure point concentrations the population could potentially be exposed to at the site.

- Application of fate and transport modeling adds an additional tier of potential uncertainty to exposure point estimates. Models cannot predict "true" exposure point concentrations at different times and places or in different media, but provide an estimate of the potential concentration under certain assumptions. Often, the assumptions used in the models are conservative to avoid underestimating potential concentrations. In addition, not all applicable processes are or can be considered (e.g., degradation, removal processes). However it is even more conservative to use current detected concentrations for exposure point concentrations for future use scenarios.

4.7.2.4 Uncertainties Associated with Toxicity Assessment. EPA-derived toxicity values are recommended to be used in risk assessments. These values are developed by applying conservative assumptions and are intended to protect even the most sensitive individuals in the populations potentially exposed. Use of these values will almost always result in overestimates of potential risk. Factors that contribute to uncertainty include:

- Use of uncertainty factors and modifying factors (MFs) in the RfD. Noncarcinogenic RfDs are primarily derived from animal toxicity studies performed at high doses to which UFs or MFs (each usually a factor of 10) are applied. This process may remove the derived dose many orders of magnitude from the dose which caused the critical effect in the study, and will most likely overestimate the site risks.

- Use of an "upper bound" cancer SF. The SF is often derived from high dose animal studies and extrapolated to low doses using extrapolation models. The 95% UCL of the slope predicted by the extrapolation model is adopted as the SF. Use of this value results in an upper bound estimate of potential risks.

- Choice of study used to derive toxicity value. The inclusion or exclusion of studies by EPA in the derivation of a toxicity value is usually made by professional judgment and affects the numerical toxicity value.

- The assumption of human sensitivity. When deriving RfDs and SFs, EPA selects a critical study (usually the animal study showing an adverse effect at the lowest exposure or intake level) as the basis for deriving the RfD or SF. EPA assumes that humans are at least as sensitive as the most sensitive animal study.

4.7.2.5 Uncertainties Associated with Risk Characterization. EPA's standard algorithms are commonly used to calculate chemical intakes and associated health risks and hazards. There are certain assumptions inherent in use of these equations that add uncertainty to the assessment.

- Assumption of additivity. Calculation of both carcinogenic risks and noncarcinogenic hazards assumes (at least as a first-line approach) additivity of toxic effects. This assumption adds uncertainty to the assessment and may result in an overestimate or underestimate of potential health risks, depending on whether synergistic or antagonistic conditions might apply.

- Omission of certain factors. The standard algorithms (without modification) do not consider certain factors, such as absorption or matrix effects. In cases where these processes are important, use of the standard algorithms without modification may result in an overestimate of potential chemical intakes.

4.7.3   Evaluation   of   Uncertainty.   Various approaches can be applied to describe the uncertainties of the assessment, ranging from descriptive to quantitative. The method selected should be consistent with the level of complexity of the assessment. It may be appropriate to conduct an in-depth quantitative evaluation of uncertainty for a detailed, complex assessment, but may not be appropriate or even needed for a screening level or relatively simple assessment. Qualitative and quantitative approaches to expressing uncertainty are discussed below.

4.7.3.1 Qualitative Evaluation. A qualitative evaluation of uncertainty is a descriptive discussion of the sources of uncertainty in an assessment, an estimation of the degree of uncertainty associated with each source (low, medium, high), and an estimate of the direction of uncertainty contributed by that source (under or overestimation). A qualitative uncertainty assessment does not provide alternate risk or hazard values, but does provide a framework in which to place the risk and hazard estimates generated in the assessment.

4.7.3.2 Quantitative Evaluation.

4.7.3.2.1   A quantitative uncertainty assessment is any type of assessment in which the uncertainty is examined quantitatively, and can take several forms. A sensitivity analysis is a form of uncertainty analysis in which the specific parameters are modified individually from which the resultant alternate risks and hazard

estimates are derived. Probabilistic approaches, such as MC simulations, are a more complex form of uncertainty analyses, and examine the effect of uncertainty contributed by a number of parameters.

4.7.3.2.2   A sensitivity analysis is a process of changing one variable while leaving the others constant and determining the effect on the output. These results are used to identify the variables that have the greatest effect on exposure. This analysis is performed in three steps:

- Define the numerical range over which each parameter varies.

- Examine the relative impact that each parameter value has on the risk and hazard estimates.

- Calculate the approximate ratio of maximum and minimum exposures obtained when range limits for a given parameter are applied to the risk algorithm.

4.7.3.2.3   A probabilistic uncertainty analysis, such as the MC simulation, examines the range of potential exposures associated with the distribution of values for input parameters of the risk algorithm. Such methods can allow the risk assessor to estimate both the uncertainty and variability associated with various parameters of a risk assessment. Uncertainty in these terms is defined as "a lack of knowledge about specific factors, parameters, or models" and variability as "observed differences attributable to true heterogeneity of diversity in a population or exposure parameter" (USEPA, 1997a).

In a probabilistic analysis, probability density functions are assigned to each parameter, then values from these distributions are selected and inserted into the exposure equation. After this process is completed a number of times, a distribution of predicted values is generated that reflects the overall uncertainty of inputs to the calculation. The results are presented graphically as the cumulative exposure probability distribution curve. In this curve, the exposure associated with the 50th percentile of the exposure may be viewed as the "average" exposure and those associated with the 90th or 99.9th percentile may be viewed as "high end" exposure.

4.7.3.2.4   An MC simulation is performed in four steps:

- Assign probability distribution functions to selected parameters in the risk algorithm.

- Develop distributions for the selected parameters (if not already available) and identify a number of randomly chosen values within that distribution.

- Apply the random input values for the parameters to the risk algorithm, and generate a number of randomly generated output values.

- Develop a cumulative probability distribution curve from the randomly generated output values.

4.7.3.2.5   A tiered approach should be used to determine the complexity, cost, and time that the project warrants for the probabilistic analysis, and whether one needs to be performed at all. Results from a traditional deterministic risk assessment should be examined prior to performing a probabilistic analysis. If the risk is close to the level of concern, the project may benefit from a probabilistic analysis. If the site clearly requires, or does not require action, further analysis is likely not necessary. The risk assessor should discuss the insight to the risk estimate that could be derived from further analysis with the risk manager as they need to be balanced with costs and time that the analysis will require.

4.7.3.2.6   A sensitivity analysis should be performed on the results of the deterministic risk assessment to determine which parameters should be focussed upon in the probabilistic assessment. To effectively utilize resources, those parameters whose uncertainty or variability has the greatest impact on the risk estimate should be assigned probability distributions in the MC simulation, other less important parameters may be held constant.

4.7.3.2.7   For more information on probabilistic analysis, including recommendations for reporting requirements, consult the *Guiding Principles for Monte Carlo Analysis* (USEPA, 1997a) or access the EPA's web site at: http://www.epa.gov/nceawww1/mcpolicy.htm. Additionally, EPA is in the process of developing RAGS Part E, *Supplemental Guidance to RAGS: The use of Probabilistic Analysis in Risk Assessment*. Several

computer-based proprietary programs are available to conduct this simulation.

CHAPTER 5

5.0 EVALUATING THE HHRA OF REMEDIAL ALTERNATIVES

5.1 INTRODUCTION

The risk assessment methodology presented in Chapters 3 and 4 focused upon the performance of the screening risk analysis used in the PA/SI, and the BRAs as appropriate for RIs. This methodology serves as the framework for all risk assessments. As mentioned earlier, risk assessments may also be performed for other aspects of site activities.

One aspect is the performance of risk assessments to support evaluations in the FS. As part of FS activities, different remedial alternatives are examined from a number of perspectives as part of the selection process. The NCP specifies nine selection criteria to be examined as part of remedial alternative evaluation: (1) protection of human health and the environment, (2) compliance with ARARs, (3) long-term effectiveness and permanence, (4) reduction of toxicity/mobility/volume through treatment, (5) short-term effectiveness, (6) implementability, (7) cost, (8) state acceptance, and (9) community acceptance. There are three risk assessment procedures that can be applied to aid in the evaluation of remedial alternatives.

The three types of risk assessments are:

- The development of RGs to be applied to site cleanup.

- The evaluation of long-term risks associated with the alternatives.

- The evaluation of short-term risks associated with implementation of the remedy.

The first type is sometimes performed as a component of the RI, but is distinguished herein because of its use in selection of remedial options. The other two types are useful in comparative evaluation of potential remedial options. They are discussed individually below.

5.2 DEVELOPMENT OF RGs

RGs are media-specific chemical concentrations that are associated with acceptable levels of chemical intake. RGs, sometimes also referred to as cleanup goals or TCLs, are considered along with other factors such as ARARs in identifying the chemical concentrations to which impacted media are to be remediated. In general, RGs are developed when the chemical-specific risks and hazards exceed acceptable levels.

RGs differ from PRGs in that site-specific factors are considered. PRGs are developed as a screening level tool prior to the performance of an RI. Conversely, RGs are developed from the site-specific BRA that was developed during the RI. See RAGS Part B (USEPA, 1991d) for a complete discussion of this process.

> **RGs Must be Developed and Applied in the Context of Exposure Area and the Exposure Point Concentration. It is Not Necessary to Remediate All Media to or Below the RG.**
>
> ---
>
> Risk assessments are based on the 95% UCL of the mean contaminant concentration. Calculation of an RG establishes a firm number to be used for cleanup. By requiring that all confimatory samples be below the RG, excessive cleanup is done and results in unnecessary cost escalation. A more realistic approach is to evaluate an exposure area, calculating concentrations that would result in a residual 95% UCL equal to the RG. The calculation includes the clean fill and the non- or minimally impacted areas. This calculation should be done as part of the RD, determining an adjusted RG. Additional information can be obtained from Bowers, et al. (1996).

RGs should be based upon all significant exposure pathways assessed in the BRA for that medium. However, since the pathways resulting in the highest degree of exposure will most greatly influence the RG, exposure pathways that have minimal contribution to overall risks can be excluded from the RG development with little or no impact. In general, if a given exposure pathway contributes less than 1 percent of the overall risks, it can be disregarded in RG development.

5.3 EVALUATION OF LONG-TERM RISKS

5.3.1 Comparative Risk Assessment of Remedial Alternatives. For a remedial alternative to be acceptable, it must be protective of human health and the environment. However, more than one alternative may meet this criteria. In these instances, an assessment of the long-term residual risks associated with both alternatives can be developed as a tool to assist in selecting an alternative. By comparing the degree to which an alternative reduces potential risks with other factors such as cost, acceptability, and effectiveness, one alternative may be preferable.

5.3.2 Risk Reduction. In addition to cost aspects, the reduction of risk offered by the alternative should be examined with respect to the risks estimated in the BRA. If the risk reduction offered is not significant, or does not address the primary risks identified in the baseline assessment, these factors should be considered in the remedy evaluation.

The reduction of risk offered by the alternative should also be examined with respect to the size of the population affected by the baseline risks or remedial alternative's reduction of risk. Although protection of all receptors is the primary goal, a modest reduction of risk for a large population may be preferable to a large reduction of risk for a small group.

5.3.3 Residual Risk. The potential risks to be addressed in a risk analysis of the alternatives are those remaining after the implementation and completion of the remedial alternatives. The calculational methodology for performing this type of the assessment is the same as for a BRA. The potential exposure pathways and receptors should also be the same as the BRA (unless temporal factors modify some of the pathways or receptors). The main factor that will change is the chemical concentration (i.e., exposure point concentration) to which the receptors may be exposed.

When developing an estimate of potential exposure point concentrations after remediation, careful consideration must be given to where remediation is to take place and where no action is anticipated. It is not uncommon for RAs to focus in some areas of a site, leaving others untouched. Therefore, estimating the

potential exposure point concentration is not as simple as assuming exposure to the RG, but will be a combination of attaining the RG in some locations, being below the RG at others, and perhaps exceeding the RG in some isolated areas where (for some other valid reason) remediation is not anticipated. The potential risks associated with different combinations of remedial alternatives can be addressed by examining each media separately, and then combining the associated risks in modular fashion.

5.4 SHORT-TERM RISKS ASSOCIATED WITH REMEDIATION

Another area in which risk assessment methodology can be applied is the evaluation of short-term risks associated with the implementation of each remedial alternative. The objective of this assessment is to evaluate whether the RA poses unacceptable potential risks to workers and other nearby receptors for each alternative evaluated in the FS.

This type of risk assessment is distinct from the BRA, as additional receptors may be exposed, and concentrations of chemicals may also differ. Additional exposure pathways may also exist. Depending on the length of time in which the remedial alternative may be carried out, short- and/or longer-term risks may need to be assessed.

This assessment focuses on the potential risks associated with the implementation and operation of the alternative. Therefore, an important component is to identify the exposure pathways potentially associated with the alternative. The risk assessor should work closely with the design engineers to identify potential for the alternative to result in exposure of workers or nearby populations. Depending on the type of alternative, exposure could occur through entrainment of soil (in the case of soil excavation), volatilization (from air stripping), or other pathways.

Once the potential exposure pathways are identified, the risk assessor needs to identify the potential degree of exposure. Remedial designers may be able to provide actual emission rates for certain alternatives. In other instances, predictive modeling may need to be applied to estimate exposure point concentrations. Once exposure factors are identified, quantitation of potential risks is calculated in the same manner as other risk assessments.

If unacceptable risks are estimated for the alternative, the use of control technologies or other management options should be examined as risk reduction measures and/or evaluation of other alternatives which may have less potential to cause short-term risks. Examples of controls include use of carbon filters on air strippers, dust suppression, use of personal protection equipment, or other controls that will reduce exposures. These factors should be weighed with other FS criteria such as cost, feasibility, schedule, risk reduction, etc., in choosing the most appropriate alternatives.

CHAPTER 6

6.0 RISK MANAGEMENT - INFORMATION NEEDED FOR DECISION-MAKING

6.1 INTRODUCTION

The NAS defines risk management as "a process of weighing policy alternatives and selecting the most appropriate regulatory action, integrating the results of risk assessment with engineering data and with social, economic and political concerns to reach a decision" (NRC, 1983). NAS has identified four key components for managing risk and resources: public participation, risk assessment, risk management, and public policy decision-makers (NRC, 1994). Risk characterization is considered the "bridge" or "interface" between risk assessment and risk management. EPA recommends that risk characterization should be clearly presented and separated from any risk management considerations. EPA (1995a) policy indicates that risk management options should be developed using risk input and should be based on consideration of all relevant factors, both scientific and non-scientific.

Consistent with NAS, USACE has developed the HTRW RMDM process. This process identifies factors to consider when making decisions, developing and recommending options, and documenting of risk management decisions (Figures 6-1, 6-2). The process establishes a framework to manage risk on a site-specific basis. It emphasizes that risk management must consider the strengths, limitations, and uncertainties inherent in the risk assessment as well as other non-risk factors. The consideration of risk is critical, since site actions are driven by statutes and regulations which explicitly require the "protection of human health and the environment."[21]

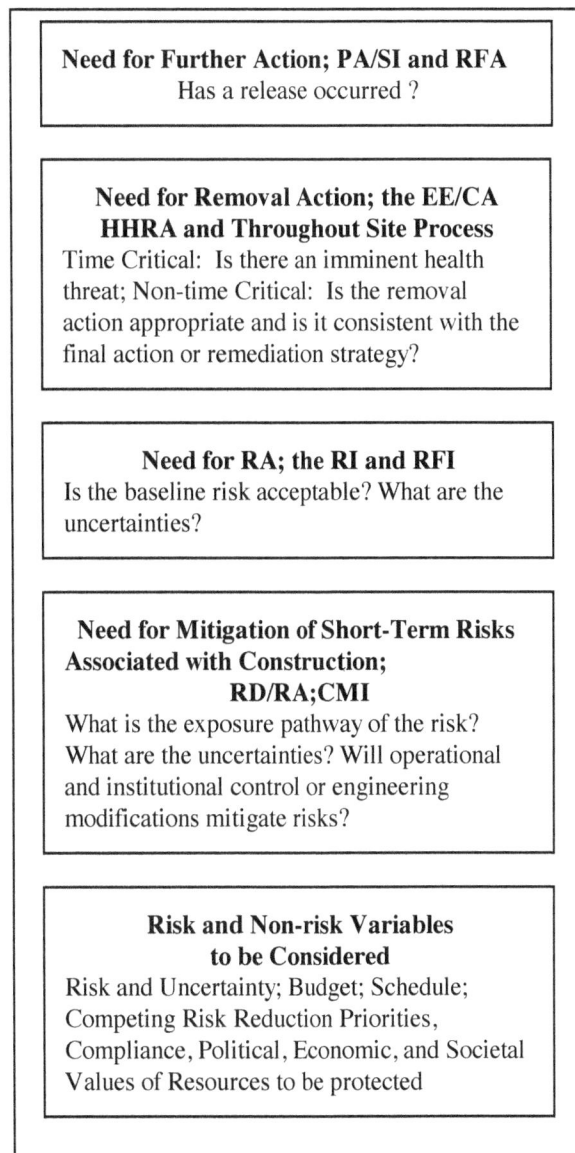

---

[21]   Examples of these requirements are 40 CFR 300.430(e)(1) of the NCP for deciding if RA is needed for a CERCLA site; RCRA Sections 3004(u), 3004(v), 3008(h), 7003 and/or 3013 for requiring corrective actions at hazardous waste TSD facilities to protect human health and the environment; and the risk-based determination for NFA (40 CFR 264.514) and selection of remedy (40 CFR 264.525) under the proposed Subpart S RCRA corrective action rules.

Need for Further Action; PA/SI and RFA
Has a release occurred ?

Need for Removal Action; the EE/CA HHRA and Throughout Site Process
Time Critical: Is there an imminent health threat; Non-time Critical: Is the removal action appropriate and is it consistent with the final action or remediation strategy?

Need for RA; the RI and RFI
Is the baseline risk acceptable? What are the uncertainties?

Need for Mitigation of Short-Term Risks Associated with Construction; RD/RA;CMI
What is the exposure pathway of the risk? What are the uncertainties? Will operational and institutional control or engineering modifications mitigate risks?

Risk and Non-risk Variables to be Considered
Risk and Uncertainty; Budget; Schedule; Competing Risk Reduction Priorities, Compliance, Political, Economic, and Societal Values of Resources to be protected

Figure 6-1. Inputs for Risk Management Decision-Making, HTRW Project Decision Diagram

| What is the project decision for the project phase? |
| Regulatory/Statutory Decision Statement |

| What are the inputs/study elements into the decision? |
| Comparison with health-based PRGs, screening risk assessment, BRA, risk analysis of alternatives, development of RAOs |

| What are the anticipated options? |
| Interim measures, removal actions, ARARs |

| What are the risk and uncertainty? |
| Reasonable maximum/high-end; average; population; and probabilistic risks |

| What are other relevant non-risk factors? |
| Risk, Uncertainty, Budget, Schedule, Competing Risk Reduction Priorities, Compliance, Political, Economic, and Societal Values of Resources to be protected, Environmental Justice, and other Stakeholders' concerns. |

| What are the options? |
| An array of potential options and their ramification on the site decision |

| What is the recommended option? |
| And the rational for the recommended option. |

| Decision by the Customer and Document Rationale for Decision |

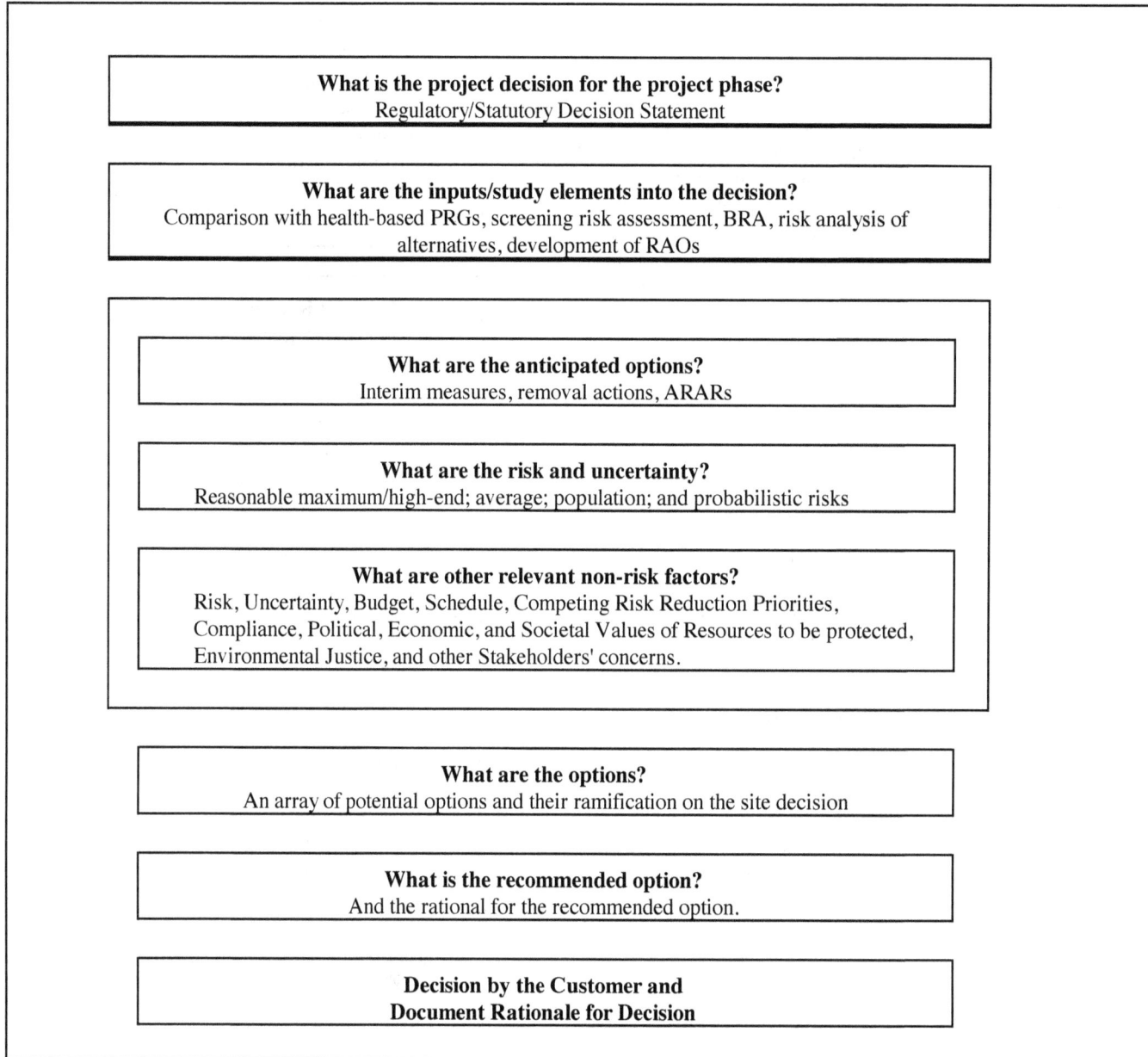

Figure 6-2. HTRW RMDM Process Flow Diagram.

Therefore, selecting the proper risk tool and collecting data to assess environmental risk is a primary responsibility of the PM and the risk assessor.

The Risk Assessment Shall be Given, at a Minimum, Equal Consideration with Other Factors in the Risk Management Decision

---

Too often, we are performing non-risk driven cleanups. Although many other factors enter into a risk management decision, the safety and health of the public, the workers, and the environment must be considered foremost. Where a sound, defensible risk assessment shows that there is little or no risk from contaminants at a site, resources should not be expended on additional study or remediation.

Additionally, data generated during the risk assessment must not be used out of context. Risk screening values must not be used as cleanup goals due to the conservative parameters used in their generation. RGs should be developed based on the calculations within the risk assessment in conjunction with the risk management decision regarding acceptable risks.

In addition to risk and uncertainty, there are many non-risk variables influencing the risk management decision. The major ones are cost, schedule, value of resources to be protected, competing risk reduction priorities among sites managed by the customer, compliance/regulatory, political, economic, and technical feasibility. A relatively sensitive political and/or economic factor to be considered is "Environmental Justice or Equity." This phrase relates to the government's initiatives to cleanup sites located in "poor and disadvantaged" areas.

The risk assessment, in conjunction with other important "non-risk" decision criteria, provides information on the need for remedial or early actions. Therefore, a clear understanding the risk assessment results and their uncertainties is essential. Informed RMDM will lead to protection of human health and the environment, cost savings, meeting the agreed schedule, political harmony, better management of resources, and other social and economic benefits. The HTRW RMDM process is

consistent with recent initiatives by various officials: Habicht (USEPA, 1992d), Denit (USEPA, 1993d), Browner (USEPA, 1995c), and DOD (1994a) that suggest the need for risk reduction based on "real world" or realistic risk assessment, cost benefit analysis, and prioritization of environmental issues.

Prior to gathering data and performing the HHRA, the PM defines the site decision for the project phase, the required study elements (types of HHRA or risk tools to be used), and the potential uncertainties associated with the outputs of the study element. Based on risk information and other considerations, the customer can select from an array of recommended risk management options. Options can include gathering additional data, recommending NFA, interim measures, or removal and/or RAs. To facilitate RMDM, the USACE PM should anticipate potential risk management options early in the project planning phase. Examples of the use of risk assessment in various project phases include:

- PA/SI or RFA: A screening risk assessment and an exposure pathways analysis may be performed to determine the need for further investigations.

- RI or RFI (prior to FS and CMS): The BRA determines the need for the RA.

- FS or CMS: Results of the BRA are used to develop RGs (i.e., the calculation of a target chemical concentration given a known target risk level or acceptable hazard).

- FS or CMS: Qualitative or quantitative risk assessments to compare and evaluate potential health impacts from the remedial alternatives. A qualitative or simple quantitative risk assessment (similar to the BRA) may be conducted to screen alternatives for their potential short-term and residual risks.

- RD (prior to conducting RA and CMI): Detailed risk analysis may be performed to determine if protective measures should be taken to minimize the impact to health and the environment during remediation. For example, a toxicity assessment may be conducted to evaluate the short-term acute, subchronic, and chronic toxicities of potential releases from the remediation process.

It is important to recognize that risk managers often make difficult decisions with considerable uncertainties in both risk and non-risk information. Therefore, a focused and balanced risk approach is recommended that recognizes the reasonable limits of uncertainty for the protection of human health and the environment as the primary consideration, along with the considerations for non-risk issues. The risk manager should clearly communicate the decision and the associated assumptions, and document the basis for the decision.

6.2 DETERMINING REQUIREMENTS FOR ACTION

The fundamental requirement associated with any HTRW response action is the "protection of human health and the environment." This requirement focuses on the acceptability of site risk or risks from the potential actions. EPA risk assessment guidelines (USEPA, 1989j), the NCP (USEPA, 1990c), and the proposed RCRA Corrective Action Rule (USEPA, 1990d) define acceptable risks of carcinogenic and noncarcinogenic effects. For carcinogens, the acceptable individual upper bound lifetime risks range from a probability of 1E-04 to 1E-06. For noncarcinogens, the acceptable hazard, expressed in terms of the sum of HQs for chemicals affecting similar organ systems or toxicological endpoints (HI), is unity. Depending on the exposure period of concern, the HQ is the average daily intake divided by the chronic or subchronic RfDs which are based on the No Observed Adverse Effects Level or the Lowest Observed Adverse Effects Level in human study or animal bioassays. Cancer risk is expressed as an individual excess lifetime risk, and is the chronic daily intake multiplied by the carcinogenic SF. Cancer risk or noncancer hazard estimates are based on the CSM specific for the site under baseline conditions, during site removal or RAs, and after remediation. Human activity patterns indicated in the CSMs are directly related to current and future land use. This paragraph presents the risk management options in key phases of the HTRW project life cycle.

6.2.1   PA/SI and RFA. The purpose of PA/SI under CERCLA and the RFA under RCRA is to identify if chemical releases have occurred, or if the site can be eliminated from further action. The PAs and RFAs are typically performed by the state, EPA, or the Federal agency, and are generally preliminary in nature. Under

some circumstances Federal agencies may perform these activities with greater depth and vigor under EO 12580. Unless good evidence exists that a site is contaminated, it is a crucial for the PM to methodically review each identified site, area of contamination, SWMU, and AOC, and decide if these units should be eliminated from the next project phase. In addition, it may be important to determine if an imminent health threat or a substantial site risk potentially exists that would require an early response action (e.g., non-time critical removal actions, interim measures, or IRA).

6.2.1.1   Actual or Potential Release/Exposure. Under the PA/SI or RFA phase, the risk management decision will be based on documented past spills and releases, the likelihood of such spills/releases, the presence of endangered or threatened species, sensitive environments or resources to be protected, and the existence of transport mechanisms that could bring the chemicals in contact with receptors.

6.2.1.2   ATSDR Health Advisories. The ATSDR performs health assessments to document or provide consultations on potential public health consequences associated with hazardous waste or Superfund sites. ATSDR representatives are located at all EPA regional offices and work cooperatively with the Superfund and RCRA staff. ATSDR involvement in the removal/emergency response program includes issuance of draft and final health advisories or consultations.

Before ATSDR health advisories are used as a basis for going forward into the next project phase or undertaking removal actions, the HTRW risk managers and PMs should contact the appropriate USACHPPM personnel for a detailed review of the health advisories to ascertain the strength and validity of the health advisories. This is recommended because the PA/SI or RFA data are quite tentative in nature, and oftentimes have not gone through a vigorous data validation process. For example, if unfiltered ground water data were used by ATSDR, and the samples had high turbidity, indicating insufficient development and purging of wells, the data should be questioned and, if feasible, new ground water data acquired to assess the need for RI, RFI, or potential removal actions.

In making risk management decisions concerning emergency response actions in this project phase, the risk managers may be put in the position of accepting data or recommendations of a lesser degree of confidence or a higher degree of uncertainty.

6.2.1.3 Risk Screening and Prioritization of Units of Concern. Initial risk screening (Chapter 3) is an important tool for ranking or prioritizing sites (OUs/SWMUs). This tool can result in substantial savings of resources, allowing the implementation of a more focused site investigation. The risk screening results are likely to provide significant inputs into the RMDM for this project phase.[22]

It is not uncommon to have tens or hundreds of "sites" or SWMUs within a site or facility boundary. Risk managers at these facilities are faced with potentially complex investigations. Rather than taking a "piece meal" approach of investigation, the list of sites or SWMUs should be pared down if possible. The risk manager may negotiate with the agencies and enter in the IAG or FFA to permit the use of an approach that "addresses the worst sites first," and at the same time, group SWMUs within the same EUs or geographical locations, as appropriate. This prioritization should result in the greatest benefit with limited available resources. Site prioritization should include the following:

- Eliminate sites or SWMUs administratively by record review, interviews with current and former workers, and ascertain whether the unit of concern meets the definition of a "SWMU."

- Conduct a site reconnaissance and group sites or SWMUs with common exposure pathways or EUs, if appropriate.

- Rank the remaining sites or groups of sites qualitatively or quantitatively based on the CSM or a screening risk analysis.

Generally, the above tools will serve well if they are objectively and uniformly applied. The use of site prioritization:

- Provides justification for NFA for low priority sites.

- Allows better resource allocation for investigation of the remaining sites.

- Helps identify potential boundaries where receptors are to be protected.

- Identifies high priority sites or SWMUs for emergency response, early actions, or accelerated cleanup or site stabilization.

The *Relative Risk Site Evaluation Primer* (DOD, 1994b) recommends evaluation based on three criteria: (1) contaminant hazard factor, (2) migration pathway factor, and (3) receptor factor. Information generated from the initial risk screening (Chapter 3) can be used as a decision-making basis using a similar site ranking process. Sites may be ranked high, medium, or low based on non-quantitative exposure pathway considerations such as the following:

1. Significant Contaminant Levels

a. High Relative Risk: Sites with complete pathways (contamination in the media is moving away from the source) or potentially complete pathways in combination with identified receptor or potential receptors.

b. Low Relative Risk: Sites with confined pathways (i.e., contaminants not likely to be

---

[22] EPA's Deputy Administrator (USEPA, 1995a,c) is concerned with the need for assuring consistency while maintaining site-specific flexibility for making remedial decisions (from site screening through final risk management decisions) across programs. EPA stresses that priority setting is reiterative throughout the decision-making process because limited resources do not permit all contamination to be addressed at once or receive the same level of regulatory oversight. EPA suggests that remediation should be prioritized to limit serious risks to human health and the environment first, and then restore sites to current and reasonably expected future uses, whenever such restorations are practicable, attainable, and cost effective. EPA further suggests that in setting cleanup goals for individual sites, we must balance our desire to achieve permanent solutions and to preserve and restore media as a resource, with growing recognition of the magnitude of the universe of contaminated media and the ability of some cleanup problems to interact with another.

released or transported) and limited potential for receptors.

c. Medium Relative Risk: Sites with characteristics not indicated in the above.

2. Moderate Contaminant Levels

a. High Relative Risk: Sites with complete pathways or potentially complete pathways in combination with identified receptors; or sites with complete pathways in combination with potential receptors.

b. Low Relative Risk: Sites with confined pathways and any receptor types (i.e., identified, potential, or limited potential), or sites with potentially complete pathways in combination with limited potential for receptors.

c. Medium Relative Risk: Sites with characteristics not indicated in 2.a and 2.b above.

3. Minimum Contaminant Levels

a. High Relative Risk: Sites with complete pathways in combination with identified receptors.

b. Medium Relative Risk: Sites with potentially complete pathways in combination with identified receptors or sites with evident pathway in combination with potential receptors.

c. Low Relative Risk: Sites with characteristics not indicated in 3.a and 3.b above.

The relative risk site ranking process may also be modified to include consideration of the degree of confidence in the relative risk rating. Sites with a low degree of confidence and a low relative risk may then be given a higher rating than sites with a high degree of confidence and a low degree of risk.

6.2.1.4 Risk Management Decisions and Options. Risk management decisions, risk information needs, risk

assessment tools to satisfy the information needs, and risk management options are presented in this section. "Non-risk" factors to be considered in the decision-making are presented in Section 6.2.4.

Risk Management Decision

- Should a site be eliminated from further investigation or included in the RI or RFI project phase?

Risk Management Options/Rationale

- **Further Evaluation Needed**

Rationale: If a site cannot be justified for NFA, further evaluation (Expanded SI; Extent of Contamination Study; RI or RFI) will be needed.

- **NFA**

Rationale:

- No knowledge of documented releases or major spills/low likelihood of spills/procedures existed to promptly cleanup all spills.

- Transport mechanisms do not exist, e.g., presence of secondary containment.

- The substances released are not expected to be present due to degradation and attenuation under the forces of the nature.

- Spills or releases have been addressed by other regulatory programs (e.g., the UST program or RCRA closure under Subpart G of 40 CFR 264 or 265).

- The unit does not meet the definition of a "SWMU."

- The unit is part of another identified unit or site which will be addressed separately.

Although risk assessment is traditionally performed in the RI or RFI project phases of HTRW response actions, risk assessment can assist the risk managers in all project phases. Results of risk assessment activities are used to answer three key questions:

- Whether or not there is a need to go forward with the next project phase.

- Whether or not early response actions (removal actions, interim measures, or IRAs) should be taken to mitigate potential risks.

- Effectiveness of the potential response action and the short-term risks associated with implementation of the removal actions.[23]

Risk Management Decision

- Should early response action be undertaken to mitigate risk?

Risk Management Options/Rationale

- **No Early Response Action**

Rationale:

- Transport mechanisms probably do not exist, e.g., presence of secondary containment.
- Low concentration of site contaminants or the levels measured probably do not pose an acute hazard, and

---

[23] Removal actions must be flexible and tailored to specific needs of each site and applicability (i.e., complexity and consistency should be used in evaluating whether non-time critical removal actions are appropriate). Examples of removal actions are: (1) sampling drums, storage tanks, lagoons, surface water, ground water and the surrounding soil and air; (2) installing security fences and providing other security measures; (3) removing and disposing of containers and contaminated debris; (4) excavating contaminated soil and debris, and restoring the site (e.g., stabilization and providing a temporary landfill cap); (5) pumping out contaminated liquids from overflowing lagoons; (6) collecting contaminants through drainage systems (e.g., french drains or skimming devices); (7) providing alternate water supplies; (8) installing decontamination devices (e.g., air strippers to remove VOCs in residential homes); and (9) evacuating threatened individuals, and providing temporary shelter or relocation for these individuals (USEPA, 1990f).

it is questionable whether the levels pose unacceptable chronic risk or hazard.

- Site contaminants are not likely to be persistent or the contaminants are relatively immobile.

- **Early Response Action**

Rationale:

- There is no current impact, but if uncontrolled, the site could pose a substantial threat or endangerment to humans or the environment. (Examples are: physical hazard, acute risk from direct contact with media of the unit or site, or effluents or contaminated media are continuously being discharged to a sensitive environment.)

- The principal threat has reasonably been identified because of the evidence of adverse impacts. In this context, the COPCs are known and the exposure pathways are judged to be complete, e.g., the exposure point or medium has been shown to contain the COPCs.

- The boundary of contamination is reasonably well defined so that removal action(s) can be readily implemented.

- The early actions are consistent with the preferred final remedy anticipated by the customer, reducing risks to human or ecological receptors, or both.

- The response action will be used to demonstrate cessation or cleanup of releases, resulting in substantial environmental gain which is the basis for early site close-out or further investigation.

- High concentration (acute hazard level) of site contaminant is found in the exposure medium.

- Highly toxic chemicals or highly persistent and bioaccumulative chemicals found on-site which may be transported off-site.

- Non-complex site (no cost recovery issue, limited exposure pathways, small area sites, etc.).

Early response actions or removal actions, consistent with the final RA, may be taken at any time to prevent, limit, or mitigate the impact of a release. To encourage early site closeout or cleanup, EPA has encouraged early response actions at sites where such actions are justified. To the extent possible the selected removal actions must contribute to the efficient performance of long-term RAs. EPA's *RCRA Corrective Action Stabilization Technologies* (USEPA, 1992n) and SACM (USEPA, 1992g) emphasizes controlling exposure and preventing further contaminant migration. While these concepts are intended to expedite site actions, risk assessment provides important information for justifying cleanup actions. The applicable risk assessment methods include:

- A screening risk analysis.

- Development of medium-specific short-term health goals for screening or comparison with modeled or site data.

- Qualitative evaluation of removal actions for their effectiveness to reduce exposure and risks.

- BRA may be appropriate for non-time-critical removal action and for complex sites (sites with multiple pathways, without ARARs, large geographic areas, and with a need for cost recovery).

In order to allow timely input into the RMDM for the removal actions or interim corrective measures, the risk assessment or risk analysis should be planned and conducted in a timely manner. If removal actions are straightforward, e.g., addressing hot spot areas or high concentration plumes, the risks associated with removal actions will then be evaluated for their potential short-term risks and hazards for the specific removal actions. The short-term risks or threats to workers and other human receptors may be based on one or more of the following:

- Air, soil, surface water, ground water (including drinking water), and food chain contamination.

- Direct (dermal) contact with contaminated media.

- Ingestion of contaminated media or inhalation of contaminated air or particulate matter.

- Fire/explosion hazard.

Early actions or accelerated cleanup can often be justified as long as the actions are consistent with the preferred site remedy. Since remedies are generally not selected until late in the FS or CMS, the customer's concept of site closeout and anticipated action is critical for deciding which types of early actions are appropriate. Based on experience gained in the Superfund program, EPA has identified certain site types where final remedies are anticipated to be the same (presumptive remedies). The current list of presumptive remedies includes:

- Municipal landfill - capping and ground water monitoring.

- Wood treatment facility - soil and ground water remediation.

- Ground water contamination with VOCs - air stripping/capture wells.

- Soil contamination with VOCs - soil vapor extraction.

6.2.1.5 Qualitative Evaluation of Response Actions for Their Effectiveness to Reduce Risks. Removal of hot spots can provide substantial improvements in the site environment. In some cases, actions can drastically reduce exposure to receptors and allow natural attenuation to further reduce the exposure point concentration. If removal actions are needed, the risk manager should request two types of risk information. First, if there is more than one removal option, what is the comparative effectiveness of the options to reduce exposure and risks? Second, what is the risk or environmental impact associated with the proposed removal action? To answer the first question, the HTRW risk assessor or risk manager provides information on how the removal option can eliminate risk or reduce the level of exposure both on-site and off-site, if contaminant migration has occurred to off-site exposure points. If substantial risk reduction can be obtained by all options, the risk manager should consider other factors, such as effectiveness, reliability, etc. To answer the second question, the project engineer estimates the destruction or treatment efficiency of the medium to be treated or disposed, and the type/quantity of wastes or contaminated debris to be generated for each potential option. This information is important if an action is likely to generate waste or damage sensitive environments in the course of the remediation.

It is important to communicate and obtain an early buy-in of the removal action from the local community. If the proposed removal actions are likely to pose unacceptable short-term risks to on-site or off-site receptors, the removal action should either be discarded or monitoring/control measures be instituted. (As discussed later, the risk assessor and HTRW TPP team members provide options for making decisions when there are divergent interests between the protection of humans and the protection of ecological receptors of concern.) The risk assessor should work with other project team members to evaluate the potential for chemical releases or habitat destruction potentially associated with a remedial option. These evaluations should be qualitative and not extensive, and can be based on a consensus of professional judgement/opinion. These individuals should recommend alternatives or precautionary/protective measures to the risk manager to mitigate any potential risks.

6.2.2   RI/RFI. The primary objective of RFI, RI, or other equivalent HTRW project phases is to determine if site contamination could pose potentially unacceptable human health or environmental risks. Determination of unacceptable risk, according to the NCP, is identified through a BRA under RME. The RCRA corrective action process is similar to Superfund for determining the need for remediation, albeit initially, the TSDF owner/operator may simply compare a specific set of SWMU data with established health-based criteria. EPA generally considers performance of a HEA to be functionally equivalent to the Superfund BRA (both human health and ecological) in the RI/FS. While a few EPA regions have developed separate guidelines for RCRA, there is a national effort underway as well. The RCRA HEA should be conducted prior to or early in the CMS to determine the need for corrective measure implementation.

If the Cumulative Site Risk Calculated in the BRA Does Not Exceed 1E-04 for Reasonable Exposure Scenarios, ARARs are Not Exceeded, and Ecological Impacts are Not Significant, No RA Should be Required.

Remediation beyond risk levels has resulted in the expenditure of excessive tax dollars. Where remediation is not justified by risk or the exceedance of ARARs, it should not be done. This point is summarized by EPA: "Where the cumulative carcinogenic site risk to an individual based on reasonable maximum exposure for both current and future land use is lass than 1E-04, and the non-carcinogenic HQ is less than 1, action generally is not warranted unless there are adverse environmental impacts." (USEPA, 1991a)

The BRA or HEA associated with the RI/RFI project phase can assist the RMDM process in the following ways:

- The BRA, performed in the RI/FS or RFI project phase, presents the degree of potential carcinogenic risks and noncarcinogenic hazards posed by the site to humans (individuals and populations), and the associated uncertainty. Risks can be estimated for the entire site, OUs, AOCs, and SWMUs.

- The results of the BRA can be used to answer the questions relating to the site decisions on: (1) whether or not there is a need to go forward with the next project phase (i.e., RD/RA needed or no action alternative); and (2) whether or not removal actions (interim corrective measures) should be implemented to mitigate potential risks, which are consistent with final action.

- If a site poses unacceptable chronic hazard or carcinogenic risk, remediation will be needed for pathways indicated in EUs. Pathways/EUs which do not pose an unacceptable risk may be eliminated from further concern. The algorithms developed in the BRA can be used in reverse to develop site-specific health-based RGs (cleanup levels) in the FS.

The above determinative factors are considered in the review of the BRA summary (and uncertainty) by the risk manager, along with other non-risk criteria in the RMDM. It should be noted that the decision could be partial, i.e., some SWMUs or sites within the facility will require remediation/removal actions while others do not.

Risk Management Decision

- Should RA or corrective measure be required based on the BRA?

Risk Management Options/Rationale

- **NFA Needed**

Rationale:

- No acute or chronic hazards of risks to humans under current and future exposure (land use) conditions/low likelihood of exposure by the receptors.

- Transport mechanisms probably do not exist.

- Low concentration of site contaminants or the levels measured probably do not pose acute and chronic hazard and carcinogenic risk.

- There is no anticipated risk of physical hazards.

- Site contaminants are not likely to be persistent or the contaminants are relatively immobile.

- Technically not feasible or impractical (e.g., dense non-aqueous phase liquid) in an aquifer not anticipated to be used for human consumption.

- **Time-Critical Emergency Response Action Needed**

Rationale:

- A high likelihood of releases and transport of site contaminants to receptors, e.g., ground water plume is migrating to onsite or offsite drinking water wells.

- A high risk of physical hazards.

- High concentration (acute hazard level) of site contaminant is found in the exposure medium.

- Highly toxic chemicals or potent carcinogens are found onsite which may be transported offsite.

- Documented unacceptable drinking water or surface water contamination, which is contacted or consumed by humans.

- **Non-Time-Critical Removal Action, Interim Corrective Measures, or Accelerated Cleanup**

Rationale:

- Principal threat to human health has been identified. If unabated, there is a potential of injury, chronic risk to humans or the environment.

- Presumptive remedies available for the identified sites or SWMUs.

- Transport mechanisms are available.

- The exposure pathway was the basis for NPL listing, or past or ongoing enforcement actions on spills or releases.

- The response action is generally consistent with the preferred site remedy, and there are no complicating factors.

- Control of migration should be taken soon, or risk the exposure of site chemicals to human receptors or valuable community resources.

- The early action will result in an incremental gain in environment benefit (including ecological), plus substantial savings in future remediation expense.

- **FS (CMS) Remediation Warranted**

Rationale:

- Unacceptable hazards and risks involving multiple chemicals and exposure pathways. If unremediated, there is a long-term threat to humans and other resources.

6-10

- Transport mechanisms are available.

- Site-specific conditions (geology or location, etc.) are unique or unusual and require detailed evaluation of remedies.

- Unusual chemicals present on site which will require bench-scale and pilot-scale studies.

- **NFA Needed Except Periodic or Continuous Monitoring**

Rationale:

- RCRA facility is operating and expected to continue for the anticipated future.

- Interim corrective measures or removal actions in place which have effectively controlled migration of site contaminants and exposure.

- Baseline risk estimates are within the acceptable range and the exposure (land use) remains in the anticipated future.

- Institutional controls are deemed adequate to control exposure.

- Toxicity of the COC, which causes the principal threat is tentative, albeit the risk or hazard has been exceeded.

- The baseline risk estimates are uncertain and there are no readily available transport media for exposure (e.g., public water supply is available in the area) or COCs are subject to natural dilution and attenuation.

6.2.3   FS/CMS and RD/RA. The FS or CMS is triggered when the baseline risk is unacceptable and remediation is needed to mitigate risks and prevent further contaminant migration. In some instances, the FS or CMS could be driven by a legal requirement to meet ARARs, although ARARs are not necessarily risk-based. The FS or CMS evaluates potential remedial alternatives according to established criteria in order to identify the appropriate remedial alternative(s). The FS or CMS can be performed for the entire site or any portion of the site that poses unacceptable risks. The results of the FS/CMS include recommendations for the risk managers or site

decision-makers, including an array of remedies for selection, RAOs, or TCLs for verification of cleanup.[24] The selected remedies/TCLs or revisions thereof will be entered into the ROD or the Part B permit.

Risk Management Decision

- **What are the RAOs?**

Risk Management Options/Rationale

The risk management decision for selection of final remedies depends substantially on the RAOs. Uses of RAOs are summarized below:

- Developed or agreed upon by the agencies prior to the FS or signing of the ROD (or modification of the RCRA permit), RAOs are used to evaluate the feasibility of candidate remediation technology in the FS.

- Initial estimation and costing of remediation (e.g., excavation and stabilization).

- Delineation of cut lines for remediation.

- For use in negotiation or final determination of specific areas, SWMUs or site-wide cleanup goals, by considering uncertainties, technology, and cost.

Before embarking on an FS, RAOs should be developed using site-specific risk information consistent with site conditions. Factors to be considered when RAOs are used as the basis for designing and implementing remediation are presented below:

6.2.3.1 RAOs Must be Based on CSM. The CSM provides the framework for the BRA and identifies the specific pathways of concern. RAOs must be able to

---

[24] For the purpose of protecting the environment, the TCLs (sometimes known as RAOs) may be the same as the environmental-based preliminary remediation levels, or they may be different. TCLs or RAOs are negotiated levels for verification of the proposed cleanup technology, practical QLs (PQLs), and uncertainties associated with the preliminary remediation levels to protect ecological resources of concern.

address these pathways and the associated risks. A refined CSM, based on the results of the BRA is paramount to the establishment of focused RAOs. The RAOs are based on preliminary remediation levels developed as the project strategy goals in Phase I of the HTRW project planning under RI/FS or RFI/CMS.

6.2.3.2 RGs Must Be Protective and Practical. RGs are performance and numerical objectives developed in the FS/CMS to assure that the remedial alternative will contribute to site remediation, restoration, and closeout/delisting. As such, they must be protective and workable. To assure protectiveness, risk-based RGs should be first derived using the BRA procedures in reverse (USEPA, 1991d). The uncertainty associated with development of the RGs should be discussed and quantified. Site decision-makers carefully consider technology, PQLs, ARARs, or TBC criteria, reference location concentrations, acceptable hazards, field or laboratory analytical uncertainties, etc., before setting the RAOs.[25]

6.2.3.3 Action Must Be Consistent with Other Project Phases. Understanding of the nature and extent of contamination, as well as the media and exposure pathways of concern, is a critical requirement for successful completion of the FS or CMS and remedy selection. Therefore, data used in the FS or CMS must interface with the RI/RFI and other previously collected site data. Inadequate data or data of poor quality misrepresent site contamination and may lead to an inadequate BRA and FS. For each exposure pathway that presents an unacceptable risk, the risk assessor and the appropriate project team members (e.g., chemist, geologist, or hydrogeologist) should review the RI data before conducting the FS. This is particularly important when the FS is performed simultaneously with the RI, based on assumptions and PA/SI or RFA data.

---

[25] Certain sites may be contaminated with natural or anthropogenic substances which pose matrix interferences and cause high sample DLs (i.e., the QLs may be higher than the environmental-based PRGs). For these sites, it may be advantageous to design a representative sampling program of the background medium to establish QLs for use as alternative RGs.

Minimal information or guidance has been developed by EPA regarding the development of RAOs for RCRA and Superfund sites. RCRA has issued the ACL Guidance based on 264.94(b) criteria and case studies (USEPA, 1988f) which may be applied to developing ACLs at the source if the acceptable ground water/surface water mixing zone concentrations and the dilution/attenuation factors are defined. Under the proposed Subpart S rule for RCRA corrective action, the state water quality criteria can be used to screen if a CMS should be conducted. Nonetheless, the key risk management issue concerning the above is that the cleanup goals must be practical and verifiable. When cleanup goals are developed to protect both humans and ecological receptors, according to Section 300.340 of the NCP, the goals must be so adjusted that both receptor types are protected.

Environmental and human health-based RAOs should be developed together and proposed to the risk manager and agencies for use in the CMS for the evaluation of remedial alternatives. It should be noted that the RAOs may have to be revised or refined based on other considerations, e.g., technology, matrix effects, target risks, uncertainties, and costs (associated with the extent of the remediation, management of remediation wastes, cost of cleanup verification analyses).

Risk Management Decision

- **What are the Remedial Alternatives or Corrective Measures?**

- **What are the Preferred or Optimal Remedial Alternatives?**

Risk Management Options/Rationale

In addition to a cost and engineering evaluation of the potential remedial alternatives, each alternative must be evaluated for its ability to reduce site risk. Among the nine criteria identified by the NCP for remedy selection, protection of human health and the environment and satisfying ARARs are considered to be the threshold (fundamental) criteria which must be met by any selected remedy. More recently, EPA has placed increased emphasis on short- and long-term reliability, cost, and stakeholders' acceptance in the overall goal to select remedies. Therefore, the assessment of residual risk (a measure of the extent of site risk reduction) is a critical task.

Screening and detailed analyses of remedial alternatives will be conducted in the FS and CMS project phase. The preferred remedial alternative will be proposed. As warranted, analysis of short-term risks to assess the need for control measures will be conducted in the RD project phase, and the control measure(s), if appropriate, will also be proposed.

In the FS, potential risk reductions associated with remedial alternatives are assessed. The relative success of one alternative over another is simply the ratio of the residual COC concentrations in the exposure medium of concern. This screening evaluation does not take into account short-term risks posed by the alternative or technology due to acute hazards, releases, or spills.

6.2.3.4 Screening Evaluation of Alternatives. This evaluation focuses on determination of short-term risks posed by the removal or remedial alternatives. The findings of this evaluation are compared among the alternatives to determine preferred remedies based on the effectiveness of the remedies to satisfy RAOs with the least impact. This screening evaluation should focus primarily on effectiveness, risk reduction, and cost.

Risk screening of alternatives should generally be qualitative or semi-quantitative. If a remedy has already been selected or is highly desirable for selection, a detailed risk analysis may not be needed. Instead, the evaluation should focus on the risk reduction of the preferred remedy, and identify any concerns or data gaps which need to be addressed. The data needed to perform this screening evaluation may come from many sources: RI or RFI data, bench scale or pilot scale treatability studies conducted for the site or from comparable sites, compatibility test, test of hazardous characteristics, field monitoring measurements, vendor's or manufacturer's information, literature values, and professional judgment.[26] Key information needed prior

to conducting the screening evaluation of remedial alternatives includes:

- Identity and quantity of emissions, effluent, byproducts, treatment residues, which may be released to the environment (during normal start-up and shut-down operations).

- Toxicity of chemical substances or COCs in the above discharges.

- Potential for dilution and attenuation.

- Existence of exposure pathways and likelihood of the pathways to be significant and complete.

- Potential for spill or releases during remediation, material handling, storage and transportation of remediation wastes.

- Potential for the causation of non-chemical environmental stressors such as destruction of critical habitat for threatened and endangered species, wetlands, or other sensitive environments.

- Temporal attributes associated with a RA which could be altered to reduce the action's impact.

- Potential release of additional COCs to the environment (e.g., re-suspension of toxic sediments during dredging, and changes of pH, redox potential, oxygen, and chemical state that may increase solubility and bioavailability).

The following are lists of qualitative evaluation criteria:

- **Risk Reduction Attributes (environmental protection, permanence, and toxicity reduction)**

- Able to remove, contain or effectively treat site COCs.

---

[26] The bench scale or pilot scale treatability studies may provide valuable information for the estimation of remedial action or residual risks. Treatability studies provide data or information on the degree of removal and/or destruction of the COCs, quantity and identity of chemicals in the emissions or effluent discharges, and potential treatment standards to be applied to satisfy RAOs. This information

is important to quantify the magnitude of risk reduction and will be useful in the comparative analysis of potential remedial alternatives.

- Able to address the exposure pathways and media of concern.

- Able to meet the RAOs and overall project strategy goals.

- **Assessment of Residual Risk Potential**

- Reasonable anticipated future land use.

- Quantity of residues or discharges to remain on site.

- Toxicological properties of the residues.

- Release potential of residues based on their fate/transport properties (e.g., log octanol/water partition coefficient, water solubilities, vapor pressure, density, etc.).

- Properties or characteristics of the environmental medium which facilitate transport (e.g., hydraulic conductivity, organic carbon contents, wind speed and direction, etc.).

- Potential for dilution and attenuation for residues released into the environment.

- The extent of, and permanence of, remediation, habitat destruction and alteration; e.g. the construction of an access road through wetlands would be considered permanent.

6.2.3.5 Detailed Analysis of Alternatives. Detailed analysis is usually conducted for the preferred remedial alternatives (or removal actions) identified in the screening evaluation described above. This detailed analysis has three objectives: (a) detailed assessment of potential short-term risk during RA, and residual risks if appropriate; (b) assess the potential for the risks to be magnified due to simultaneous implementation of this and other preferred alternatives; and (c) identify potential risk mitigation measures for the preferred remedies. The findings of these tasks are presented for final selection of remedies prior to ROD sign-off or RCRA Part B permit modification. All preferred remedies or options should satisfy RGs and should pose minimum health and environmental impact.

This evaluation may be qualitative, semi-quantitative, or quantitative. If the analysis is quantitative, procedures and approaches similar to the BRA may be followed. The *Air/Superfund National Technical Guidance Study Series* (USEPA, 1989a, 1990d, 1992o, 1993c, and 1995b) includes documents providing guidance for rapid assessment of exposure and risk. For example, guidance on determining the volume of soil particulates generated during excavation is provided in *Estimation of Air Impacts for the Excavation of Contaminated Soil* (USEPA, 1992b). The data sources used to perform this risk analysis task should be similar to those identified for the screening evaluation of remedial alternatives. Although it is conceivable that the level of effort required for this analysis may be high (particularly if the same analysis has to be performed for a number of preferred remedies), it is anticipated that the documentation and report writing will be focused and streamlined.

The report should focus on the risk analysis approaches, sources of data, findings/recommendations for risk mitigation measures, and appendices. Key factors or criteria to be considered in the screening evaluation of remedial alternatives are:

- The criteria or considerations in the assessment of short-term and residual risks are substantially similar to those identified for the screening evaluation of remedial alternatives. The key difference may be additional use of quantitative data input into the risk calculations, e.g., sediment transport modeling to evaluate the potential for migration of toxic sediment, amount of discharges or emissions, dilution/attenuation or atmospheric dispersion factors, exposure frequency, duration, and other activity patterns which could impact existing vegetation and wild life in time and space.

- Time required and extent of recovery from exposure to the COCs.

- The potential for fire, explosion, spill, and release of COCs from management practice of excavated or dredged materials should remain qualitative or semi-quantitative. Fault-tree (engineering) analysis for accidental events may be attempted under special circumstances (e.g., to address public comments or if demanded by citizens during public hearing of the proposed remedies).

6.2.3.6 Risks from Simultaneous Implementation of Preferred Remedies.

- Common exposure pathways for effluent or discharges from remedies.

- Period of exposure to receptors via the common locations, time, and pathways.

- Sensitive environments and other threatened or sensitive wildlife or aquatic populations.

- Risk estimates or characterization results.

- Toxicological evaluation for the validity of additivity of risk (e.g., under the Quotient Method), based on literature review, mode of action, and common target organs, etc.

- Qualitative or quantitative assessment of potential short-term or residual risks.

Short-Term Risks Associated with Construction; the Design Risk Analysis

All removal or remedial alternatives have a potential to pose short-term risks to on-site mitigation workers, ecological receptors, and off-site humans. Risks may be associated with vapors, airborne particles, treatment effluent, resuspension of sediment resulting in an increase in the total suspended solids or siltation of substrate for macroinvertebrates, and residues generated during operation of the remedial alternative. Therefore, all the alternatives should be reviewed for their short-term risks in conjunction with data from their bench scale or pilot scale treatability studies or data from implementation of the remedy at comparable sites. The risk assessor should estimate the period of recovery from these short-term insults and determine if biological or chemical monitoring of the effects of remediation activities should be implemented. For all practical purposes, risk may remain upon completion of the RA (residual risk).

Long-Term Risks Associated with Alternatives; the Residual Risks

Unless all sources of contamination are removed or isolated, there will be residual risks at the site upon

completion of the RA. The COC residuals could either remain or be quickly degraded, depending on the COC's physical and chemical properties. The level of residual risk will depend on the effectiveness of the remedy in containing, treating or removing site contaminants, and the quantity, and physical, chemical, and toxicological characteristics of residues or byproducts remaining at the site. Site COCs which remain on-site after the RA should be assessed for their potential risks.

This evaluation step focuses on a risk reduction assessment to determine if a potential remedial alternative is able to meet the RAOs, and an assessment of residual risk potential. The findings of these tasks are compared among the alternatives to determine an array of preferred remedies based on the effectiveness of the remedies to satisfy RAOs with the least long-term health and environmental impacts.

RA/Residual Risks vs. Baseline Risk

There are notable differences between RA/residual risks and the baseline risk. The key difference is that baseline risk refers to the potential risk to receptors under the "no remedial action" alternative, and RA and residual risks refer to short-term risks during RA and long-term risks which may remain after completion of the RA, respectively. Residual risk may be considered comparable to baseline risk after remediation, since in both cases the risks are chronic or subchronic in nature. RA risks are generally short-term (acute or subchronic) risks.[27]

6.2.4 Non-Risk Issues or Criteria as Determining Factors for Actions. The NCP recognizes that it is not possible to achieve zero risk in environmental cleanup; therefore, the approach taken by Superfund is to accept

---

[27] One exception (i.e., remedial risk which is long-term) is a pump-and-treat remedy of ground water to meet MCLs for organics which pose a threat to human health but not ecological receptors. If the effluent is discharged to a surface water body and happens to contain trace elements at high levels (or other COCs not reduced by the treatment process), then an exposure route to environmental receptors may remain which is not addressed by the BRA, and which will exist for the operational lifetime of the remedy.

non-zero risk and return the site to its best current use (not to conditions of a pre-industrialization era). Under RCRA, the preamble to the proposed Subpart S recognizes that cleanup beyond the current industrial land use should be justified. This section presents and discusses the non-risk factors, and recommends a balanced approach for resolution of issues to enable quality RMDM. These factors can be categorized into scientific and non-scientific factors, as explained below.

6.2.4.1 Scientific Factors. The scientific factors, including engineering design and feasibility, should be considered in RMDM. These factors focus on technology transfer (realistic performance of the technology), duration of protection, and FS data uncertainties. These factors will influence the decision whether or not to proceed with selection of a particular remedy. They are detailed below:

Technology Transfer. This factor concerns the treatability of the contaminated debris or media by a preferred technology or early action. Although the recommended technology may appear attractive, a number of problems must be overcome before actual selection or implementation of the action. The following are a few examples:

- Scale up.

- Downtime and maintenance (including supplies).

- Ownership/control.

- Throughput to meet the required completion schedule.

- Skills required or training requirements.

- QLs and DLs.

- Space requirements for the remediation process and management of remediation wastes.

Duration of Protection. This factor concerns the duration of the removal or remedial technology designed to treat or address site risk. This factor is particularly important for site radionuclides or non-aqueous phase liquid compounds in the aquifer. The maintenance or replacement of barriers or equipment is also a primary

concern for this factor. Although a technology or alternative is effective, its effectiveness may not last long if there is no source control or contamination from off-site sources is not controlled

Data Uncertainty. This factor considers reliability and uncertainty of certain site or FS data for use in selecting a remedy, or for determining whether NFA is appropriate. Uncertainty in the following data may also impact the risk analyses or BRA results:

- Adequacy of bench-scale or pilot-scale treatability data.

- Data uncertainties (volume, matrices, site geology/hydrogeology).

- Field data and modeling data.

- Overall uncertainty of the source of site contamination.

6.2.4.2 Non-Scientific Factors. Non-scientific factors should also be considered in RMDM because some of these factors are key to a successful site remediation. Most of these factors are internal, but can also be external. Examples of these factors are enforcement, compliance, schedule, budget, competing risk reduction priorities, community inputs, and societal/economic value of the resources to be protected. These factors will influence the decision on whether or not certain removal or RAs should be taken, or on which remedies are to be selected. These factors are detailed below.

Enforcement and Compliance. Certain courses of action (including risk management decisions) have been agreed upon early in the process and have been incorporated in the IAG or FFA. This is particularly germane to some earlier HTRW sites. Nonetheless, the requirements specified in the enforcement documents or administrative order of consent, IAG, FFA should be followed by the risk manager or PM with few exceptions. When risk-related factors or other non-risk factors are over-arching, the risk manager should then raise this issue to higher echelon or to the legal department for further action or negotiation.

Competing Risk Reduction Priorities. Although related to risk, this factor represents the competing interest among programs or within the project for a limited source of funding to perform risk reduction activities. Since it is likely that not all sites will be cleaned up at an equal pace, the planning and execution of environmental restoration among these units should follow a prioritization scheme. However, the scheme developed according to risk may not be the same according to the customer, the base commander, or the agencies. The risk manager or PM must seek common ground to resolve this issue so that resources can be expended to produce incremental environmental benefits.

Schedule and Budget. These factors usually go together because the more protracted the project life, the more resources the project will demand. While each PM would like to comply with risk-based considerations with little margin of error, the PM may have no choice but to make risk management decisions with larger uncertainties than he or she would prefer, due to schedule and budget constraints.

Community Input. Opportunity for the stakeholders or community to provide input into the permit modification is provided when primary documents are prepared, i.e., RFI Work Plan, RFI/CMS reports, the proposed remedies, and the CMI Work Plan. Superfund also provides similar opportunities for public participation. To be successful in site remediation and closeout, the risk managers must be able to communicate risks effectively in plain and clear language without bias. Early planning and solicitation of community input is essential to democratization of RMDM. Some of the following issues may be of concern to the communities:

- Ineffective communication of risks and uncertainties.

- Lack of action (some action is preferred to no action).

- Not in my backyard (off-site transportation of contaminated soil, debris or sediment should avoid residential neighborhoods).

- Any treatment effluent or discharge is unacceptable (on-site incineration is seldom a preferred option except for mobile incinerators, in certain instances).

- The remedy should not impede economic growth or diminish current economic and recreational value of resources to be protected.

- Cleanup will improve the quality of life and increase property values or restoration of recreational or economic resources.

Societal/Economic Value of the Resources to be Protected. This non-risk factor concerns the community sentiment on how fast or in what manner the resources impacted by site contaminants should be restored. These resources may include surface water bodies, wildlife, and game animals. Most communities would like to see impacted resources restored to original use, however, this can be difficult to achieve. Some communities may be willing to accept natural attenuation or no action options for impacted surface water bodies, given the opportunity to examine the pros and cons of all options. Therefore, it is recommended that the risk manager execute a community relations plan in earnest in order to solicit the citizens' input on the risk reduction approach and issues of concern. Key community spokespersons may also be appointed to the site action committee to facilitate such dialogue and communication.

6.2.4.3 A Balanced Approach. In conclusion, the risk manager should consider all risk and non-risk criteria before making risk management site decisions. Due to uncertainties associated with risk assessment or analysis, the decision-maker must review risk findings and the underlying uncertainties, and consider other non-risk factors in the overall risk management equation. When making risk management decisions, the risk manager should keep an open mind regarding the approaches to meet the project objective. In order to make informed site decisions, the risk assessor must present risk estimates in an unbiased manner. With an understanding of the volume of contaminants of concern, significance and relevance of the effects and potentially impacted receptors, fate/transport properties of the COCs, and completeness of the exposure pathways, the risk manager, PM, and stakeholders will be better equipped to make informed decisions. These decisions should be consistent with the overall site strategy, which is developed early in the project planning phase, and which may evolve throughout the project.

6.3 DESIGN CONSIDERATIONS

Risk assessment methodology can be an important tool in the design phase of CERCLA RAs or RCRA corrective measure implementation. During the early phase of RD/RA or CMI, risk assessment results can help determine: 1) whether the selected remedy can be implemented without posing an unacceptable short-term risk or residual risk; and 2) control measures (operational or engineering) to mitigate site risks and to assure compliance with ARARs, TBC requirements, and permit conditions. The risk and safety hazard information will be evaluated by the site decision-makers, along with information concerning design criteria, performance goals, monitoring/compliance requirements prior to making risk management decisions regarding the above questions. Further, the decision-makers consider potential requirements such as ARARs and TBCs in determining design changes or control measures.

This section addresses the above issues, i.e., risk management considerations in RD, compliance with ARARs, including the CAA, CWA, ESA, and other major environmental statutes, and control measures required to mitigate risks.

6.3.1 Potential Risk Mitigation Measures.

Engineering Control - Where appropriate (when short-term risks are determined to be unacceptable), engineering controls should be recommended by the design engineer with inputs from the risk assessor, ecologist, compliance specialist, and the air modeler. Examples of these control measures include:

- VOC and Semi-Volatile Organic Compounds (SVOC) emissions - activated carbon canisters, after burners, or flaring, prior to venting.

- Metals and SVOC airborne particles - wetting of work areas; particulate filter/bag house, wet scrubber, or electrostatic precipitator (for thermal treatment devices or incinerators).

- Fugitive emissions - monitoring of valves, pipe joints, and vessel openings; and barrier/enclosure of work areas (e.g., a can or shield over the auger stem).

- Neutralization or chemical deactivation of effluent (continuous process or batch).

- Use of remote control vehicle for handling, opening, or cutting of drums containing explosive or highly reactive or toxic substances.

6.3.1.1 Operational Control. Where appropriate, administrative control measures (procedural and operational) safeguards should be recommended by the PM, design engineer, or field supervisor during RA, with inputs from the risk assessor and other relevant technical and compliance specialists. Examples of these control measures include:

- Establish short-term trigger levels which will require work stoppage or upgrade of the remediation procedures (e.g., dredging of toxic sediments). Either biological or chemical indicators, or their combination could be used as the trigger levels. These levels should be developed in the RD/RA or CMI project phase by the risk assessor and other technical specialists, including the modeler.

- Consistent with the above trigger or acute concern levels, evaluate on-site performance with field equipment to assure adequate remediation.

- Afford the proper protection of sensitive environments by careful planning and positioning of staging area, storage or management of remediation wastes, selection of equipment with low load bearing, and season or time period when the remediation should be completed.

- Establish a zone of decontamination and proper management of effluent or waste generated from this zone.

- Secure and control access to areas where RAs are being implemented at all time.

6.3.1.2 Institutional Control. Institutional controls are particularly pertinent for remedies which involve containment, on-site disposal of wastes, or wetlands remediation. Institutional controls should be recommended by the customer, PM, and other site decision-makers. Examples of these control measures include:

- Recording land use restrictions in the deeds (deed restrictions) for future use of certain parcels or areas where hazardous substances or wastes are contained.

- Erection of placards, labels, and markers which communicate areas where human exposure may pose short-term or residual risks.

- Security fences and barriers.

6.3.2   Risk Management; Degree of Protectiveness. Not only should a selected RA (corrective measure) be able to meet balancing criteria, the RA must be protective, i.e., in terms of reducing site risks. In designing a selected remedy, the site decision-makers may face operational or engineering issues which are likely to require risk management decisions. For example, if a detailed analysis of a selected remedy reveals potential short-term or residual risks, the decision-makers must decide to what extent and with what control measures are necessary to abate the risk. Inputs from the risk assessor will be needed to help make informed risk management decisions. The following are examples of key risk management considerations for designing an effective remediation strategy:

- **Acceptability of control measures**. There are potential operational (procedural) or engineering control measures to address the short-term risks. The risk assessor, in coordination with the design engineer, expert ecologist(s)/advisory panel, and other project team members, assesses the effectiveness of any proposed control measures.

- **Removal of control measures**. Before a control measure is implemented, the decision on the minimum performance and when to stop requiring the control measure has to be addressed. This is particularly important if control measures are costly to implement and maintain.

- **Effectiveness of the remediation**. Remediation should effectively address on-site contamination if there is an continuing off-site (regional) source. This consideration is particularly important for ground water and sediment contamination remediation. This regional source control strategy should not be confused with the identification of Potentially Responsible Parties since some of the

discharges could be a permitted activity. Nonetheless, this issue has to be resolved if the RAOs are risk-based and do not consider off-site influences or contribution to the contaminants requiring remediation.   Off-site source control and containment, waste minimization, and closure issues should be raised by the risk manager to the agencies, USACE customers, and higher echelon.

- **BRAC**. With BRAC, the land use of closed defense facilities may not be indefinitely controlled and the legislation governing BRAC holds the U.S. government responsible for future cleanup of contamination caused by government activities. Cleanup criteria and long-term remedies should take land use into consideration for implementation of an effective site closeout strategy.   For example, conversion of a military base into a state park or refuge area will require different cleanup objectives than cleanup to the level acceptable for industrial/commercial usage. This issue should be addressed early in the site strategy development phase with input from customers, local re-development commissions, state, and other stakeholders.

- **Verification of cleanup**. The risk management decision concerning verification of cleanup, i.e., the numerical value of the RAO, should be based on a combination of factors: risk, uncertainty, statistics, analytical DLs/matrices, and costs. Although RAOs have been negotiated or determined in the ROD, the sampling method and statistical requirements must be clearly articulated before design and implementation of the corrective measures or remedial alternatives.

Risk management decisions during the design phase of a CERCLA or RCRA remediation should be flexible, considering the uncertainty in the risk assessment results, acceptable risk range, confidence level of toxicity data or criteria to support the assessment, engineering feasibility, reliability of the measures (operational changes vs. pollution control equipment), state and community acceptance, and cost. It is recommended that risk managers and site decision-makers request input from all members of the project team for pros and cons of proposed control measures to address the short-term risks.

APPENDIX A
REFERENCES

A.1 REQUIRED PUBLICATIONS.
References listed below have been cited in the text.

Executive Order (EO) 11990 (Presidential Document), 1977 (May). Protection of Wetlands.

EO 12088. October 13, 1978. Federal Compliance with Pollution Control Standards. 43 FR 47707.

EO 12498. January 8, 1985. Regulatory Planning Process. 50 FR 1036.

EO 12580. January 29, 1987. Superfund Implementation. 52 FR 2923.

EO 12777, 1991. Implementation of Section 311 of the Federal Water Pollution Control Act of October 18, 1972 and the Oil Pollution Act of 1990.

U.S. Department of Defense (DOD), 1973 (May). Directive 5100.50. Protection and Enhancement of Environmental Quality.

DOD, 1977a. Directive 5030.41. *Oil and Hazardous Substances Pollution Prevention and Contingency Program.*

DOD, 1977b (August). Directive 4120.14. *Environmental Pollution, Prevention, Control, and Abatement.*

DOD, 1978. Directive 6230.1. *Safe Drinking Water.*

DOD, 1979 (July). Directive 6050.1. *Environmental Effects in the United States of Department of Defense Actions.*

DOD, 1993. Memorandum from the Deputy Secretary of Defense to Addressees; Subj: Fast Track Cleanup at Closing Installations. Washington, D.C. September 9.

DOD, 1994a. Memorandum from the Deputy Secretary of Defense to Addressees; Subj: Fast Track Cleanup - Finding of Suitability to Transfer (FOST) for BRAC Property. Washington, D.C.

DOD, 1994b. *Relative Risk Site Evaluation Primer.* Interim Edition. Washington, D.C.

U.S. Army (USA), 1991 (February). Department of the Army Pamphlet (DA Pam) 40-578. *Health Risk Assessment Guidance for Installation Restoration Program and Formerly Used Defense Sites.*

USA. Army Regulation (AR) 200-1. *Environmental Protection and Enhancement.*

U.S. Army Corps of Engineers (USACE). EM 200-1-1. *Validation of Analytical Chemistry Laboratories.*

USACE. EM 200-1-2. *Technical Project Planning (TPP) Process.*

USACE. EM 200-1-3. *Requirements for the Preparation of Sampling and Analysis Plans.*

USACE. EM 200-1-4. *Risk Assessment Handbook: Volume II - Environmental Evaluation.*

USACE. EM 200-1-6. *Chemical Quality Assurance for HTRW Projects.*

USACE, 1996a (January 17). *USACE HTRW Management Plan.*

USACE, 1992. *Environmental Compliance Assessment System (ECAS) Assessment Protocols.* Construction Engineering Research Laboratories (CECER).

U.S. Environmental Protection Agency (USEPA), 1985 (Feb). EPA/600/8-85/002. *Rapid Assessment of Exposure to Particulate Emissions from Surface Contamination Sites.*

USEPA, 1986. EPA/530-SW-86-053. *RCRA Facility Assessment Guidance.* Office of Solid Waste/Waste Management Division.

USEPA, 1987a (July). EPA/600/8-87/042. *Selection Criteria for Mathematical Models Used in Exposure Assessments: Surface Water Models.*

USEPA 1987b (July). EPA/530/SW-87/017. *Alternate Concentration Limit Guidance, Part 1. ACL Policy and*

Information Requirements. Office of Solid Waste/Waste Management Division.

USEPA, 1988a. EPA/530-SW-88-028. *RCRA Corrective Action Plan*.

USEPA, 1988b. *Enforcement Actions under RCRA and CERCLA at Federal Facilities*.

USEPA, 1988c. *Evaluation Process for Achieving Federal Facility Compliance*.

USEPA, 1988d (April). EPA/540/1-88/001. OSWER Directive 9285.5-1. *Superfund Exposure Assessment Manual*. Office of Emergency and Remedial Response.

USEPA, 1988e (May). EPA/600/8-88/075. *Selection Criteria for Mathematical Models Used in Exposure Assessments: Ground-Water Models*.

USEPA, 1988f (May). EPA/530/SW-87/017. *Alternate Concentration Limit Guidance Based on 264.94(b) Criteria Case Studies*. Office of Solid Waste/Waste Management Division.

USEPA, 1988g (June). EPA/530-SW-88-029. *RCRA Corrective Action Interim Measures Guidance*. Interim Final. Office of Solid Waste and Emergency Response.

USEPA, 1988h (September). EPA-450/4-88-009. *A Workbook of Screening Techniques for Assessing Impacts of Toxic Air Pollutants*.

USEPA, 1988i (October). EPA/540/G-89/004. OSWER Directive 9355.3-01. *Guidance for Conducting Remedial Investigations and Feasibility Studies Under CERCLA*. Office of Emergency and Remedial Response.

USEPA, 1988j (November). *Federal Facilities Compliance Strategy*. Office of Federal Activities, Office of External Affairs.

USEPA, 1989a (January). EPA-450/1-89-003. *Air/Superfund National Technical Guidance Study Series: Volume III - Estimation of Air Emissions from Cleanup Activities at Superfund Sites*. Office of Air Quality Planning and Standards.

USEPA, 1989b (March). EPA/540/1-89/002. *Risk Assessment Guidance for Superfund, Volume I (RAGS): Human Health Evaluation Manual*. Interim Final. Office of Emergency and Remedial Response.

USEPA, 1989c (March). EPA/600/3-89/013. *Ecological Assessment of Hazardous Waste Sites: A Field and Laboratory Reference*. Office of Research and Development.

USEPA, 1989d (March). EPA/625/3-89/-16. *Interim Procedures for Estimating Risks Associated with Exposures to Mixtures of Chlorinated Dibenzo-p-Dioxins and -Dibenzofurans (CDDs and CDFs) and 1989 Update*.

USEPA, 1989e (17 April). EPA Order 5360.1. *Policy and Program Requirements to Implement the Quality Assurance Program*. Office of the Administrator.

USEPA, 1989f (May). *RCRA Facility Investigation (RFI) Guidance*. *Vol. I*. Interim Final. Office of Solid Waste, EPA.

USEPA, 1989g (August). EPA/540/G-89/006. *CERCLA Compliance With Other Laws Manual - Draft Guidance*. Office of Emergency and Remedial Response.

USEPA, 1989h (August). EPA/540/G-89/009. OSWER Directive 9234.1-02. *CERCLA Compliance With Other Laws Manual - Part II*. Office of Emergency and Remedial Response.

USEPA, 1989i (August). *Federal Facilities Negotiation Policy*. Office of Emergency and Remedial Response. OSWER Directive 9992.3.

USEPA, 1989j (December). EPA/540/1-89/002. *Risk Assessment Guidance for Superfund: Vol. 1 - Human Health Evaluation Manual (Part A)*. Office of Emergency and Remedial Response.

USEPA, 1990a (January). OSWER Directive 9992.4. *Federal Facilities Hazardous Waste Compliance Manual*. Office of Waste Programs Enforcement.

USEPA, 1990b (January). EPA/600/6-90/003. *Methodology for Assessing Health Risks Associated with Indirect Exposure to Combustor Emissions.*

USEPA, 1990c (May). *National Oil and Hazardous Substances Pollution Contingency Plan.* Final Rule. Office of Solid Waste and Emergency Response. 55 FR 8660.

USEPA, 1990d (July). *Corrective Action for Solid Waste Management Units at Hazardous Waste Management Facilities.* Proposed Rule. Office of Solid Waste. 55 FR 30798.

USEPA, 1990e (August). EPA-450/1-89-002a. *Air/Superfund National Technical Guidance Study Series: Volume II - Estimation of Baseline Air Emissions at Superfund Sites.* Office of Air Quality Planning and Standards.

USEPA, 1990f (September). EPA/540/8-89/014. *Superfund Emergency Response Actions - a Summary of Federally Funded Removals, Fourth Annual Report, Fiscal Year 1989.* Office of Research and Development.

USEPA, 1991a. OSWER Directive 9355.0-30. *Role of the Baseline Risk Assessment in Superfund Remedy Selection Decisions.* Memorandum from Don R. Clay, the Assistant Administrator to Regional Division Directors.

USEPA, 1991b (March). Timothy Fields, Jr. Memo, OSWER Directive 9285.6-03. *Human Health Evaluation Manual, Supplemental Guidance: "Standard Default Exposure Factors."*

USEPA, 1991c (September). OSWER Directive 9345.0-01A. *Guidance for Performing Preliminary Assessments Under CERCLA.* Hazardous Site Evaluation Division, Office of Emergency and Remedial Response.

USEPA, 1991d (December). OSWER Directive 9285.7-01B. *Human Health Evaluation Manual, Part B: "Development of Risk-Based Preliminary Remediation Goals.* Office of Emergency and Remedial Response.

USEPA, 1991e (December). OSWER Directive 9285.7-01C. *Human Health Evaluation Manual, Part C: "Risk Evaluation of Remedial Alternatives".* Office of Emergency and Remedial Response.

USEPA, 1992a. *Hazard Ranking System Guidance.* Interim Final.

USEPA, 1992b. EPA-450/1-92-004. *Estimation of Air Impacts for the Excavation of Contaminated Soil.*

USEPA, 1992c (January). EPA/600/8-91/011B. *Dermal Exposure Assessment: Principles and Applications.* Office of Health and Environmental Assessment.

USEPA, 1992d (February 26). *Guidance on Risk Characterization for Risk Managers and Risk Assessors.* Memorandum from F. Henry Habicht, Deputy Administrator.

USEPA, 1992e (March). OERR 9200.6-303. *Health Effects Assessment Summary Tables; Annual FY 92.* Environmental Criteria and Assessment Office. Prepared for Office of Emergency and Remedial Response.

USEPA, 1992f (March). EPA/600/R-92/047. *Reference Guide to Odor thresholds for Hazardous Air Pollutants Listed in the Clean Air Act Amendments of 1990.*

USEPA, 1992g (April 7). OSWER Directive 9203.1-01. *Superfund Accelerated Cleanup Model (SACM).* Office of Solid Waste and Emergency Response. (Also see OSWER Directive 9203-1-03, July 7, 1992, *Guidance on the Implementation of the SACM under CERCLA*).

USEPA, 1992h (April). OSWER Directive 9285.7-09A. *Guidance for Data Useability in Risk Assessment (Part A).* Final report. Office of Emergency and Remedial Response.

USEPA, 1992i (May 29). *Guidelines for Exposure Assessment.* 57 FR 22888.

USEPA, 1992j (May). OSWER Directive 9285.7-081. *Supplemental Guidance to RAGS: Calculating the Concentration Term.* Office of Solid Waste and Emergency Response.

USEPA, 1992k (May). Publication No. 9285.7-09B. PB92-963362. *Guidance for Data Useability in Risk Assessment.* Part B.

USEPA, 1992l (September). OSWER Directive 9234.2-22FS. *ARARs Fact Sheet - Compliance With Clean Air Acts and Associated Air Quality Requirements.* Office of Emergency and Remedial Response.

USEPA, 1992m (September). OSWER Directive 9345.0-05. *Guidance for Performing Site Inspections Under CERCLA.* Office of Emergency and Remedial Response.

USEPA, 1992n (October). EPA/625/R-92/014. *RCRA Corrective Action Stabilization Technologies - Proceedings.* Office of Research and Development.

USEPA, 1992o (November). EPA-450/1-89-001a. *Air/Superfund National Technical Guidance Study Series: Volume I - Overview of Air Pathway Assessments for Superfund Sites (Revised).* Office of Air Quality Planning and Standards.

USEPA, 1993a (February 16). *Corrective Action Management Units and Temporary Units; Corrective Action Provisions; Final Rule.* Office of Solid Waste. 58 FR 8658.

USEPA, 1993b (March). ECAO-CIN-842. *Provisional Guidance for Quantitative Risk Assessment of Polycyclic Aromatic Hydrocarbons*, Final Draft.

USEPA, 1993c (May). EPA-451/R-93-007. *Air/Superfund National Technical Guidance Study Series: Volume IV - Guidance for Ambient Air Monitoring at Superfund Sites.* Office of Air Quality Planning and Standards.

USEPA, 1993d. Initiative by EPA Official (Denit).

USEPA, 1994a. *Laboratory Data Validation, Functional Guidelines for Evaluating Organics Analyses.* EPA Hazardous Site Evaluation Division, Data Review Work Group.

USEPA, 1994b (July 1). *Laboratory Data Validation, Functional Guidelines for Evaluating Inorganics Analyses.* EPA Hazardous Site Evaluation Division, Data Review Work Group. July 1.

USEPA, 1994c (August). OSWER Directive #9355.4-12. *Revised Interim Soil Lead Guidance for CERCLA Sites and RCRA Corrective Action Facilities.* Office of Solid Waste and Emergency Response.

USEPA, 1995a (February). *Guidance for Risk Characterization.* (Attachment to C. Browner Memorandum, 1995). Science Policy Council. February.

USEPA, 1995b (February). EPA-454/R-95-003. *Air/Superfund National Technical Guidance Study Series: Volume V - Procedures for Air Dispersion Modeling at Superfund Sites.* Office of Air Quality Planning and Standards.

USEPA, 1995c (March 21). "EPA Risk Characterization Program" and "Policy for Risk Characterization at USEPA." Memo from C. Browner.

USEPA, 1995d (November). *Supplemental Guidance to RAGS: Region IV Bulletins, Human Health Risk Assessment. Bulletin No. 1 - Data Collection and Evaluation.*

USEPA, 1996a (May). EPA/540/R-95/128. *Soil Screening Guidance: Technical Background Document.* Washington, D.C.

USEPA, 1996b (May). EPA/540/R-96/018. *Soil Screening Guidance: User's Guide.*

USEPA, 1996c (December). *Recommendations of the Technical Review Workgroup for Lead for an Interim Approach to Assessing Risks Associated with Adult Exposures to Lead in Soil.*

USEPA, 1997a (March). EPA/630/R-97/001. *Guiding Principles for Monte Carlo Analysis.*

USEPA, 1997b (June 5). *Ecological Risk Assessment Guidance for Superfund: Process for Designing and Conducting Ecological Risk Assessments*, Interim Final

USEPA, 1997c (August). EPA/600/P-95/002Fa, b, and c. *Exposure Factors Handbook.*

USEPA, 1998a (January). OSWER Directive 9285.7-01D. EPA/540/R-97/033. *Risk Assessment Guidance for Superfund, Volume I: Human Health Evaluation Manual, Part D.*

USEPA, 1998b (April). EPA/630/R-95/002F. *Guidelines for Ecological Risk Assessment*, Final.

U.S. Air Force (USAF), 1989. *The Air Force Installation Restoration Program Management Guidance*. Washington, D.C.

USAF, 1991. *USAF Hazardous Waste Management Policy*. HQUSAF, Washington, DC.

USAF, 1992. *FY 93/94/95 DERA Eligibility and Programming Guidance*. HQ USAF. Washington, D.C.

U.S. Department of the Navy (DON), 1992. *The Navy and Marine Corps Installation Restoration (IR) Program Manual*. Chief of Naval Operations (CNO). Washington, D.C.

DON, 1994 (July). OPNAVINST 5090.1B. *Environmental and Natural Resources Program Manual*. Office of the Chief of Naval Operations.

U.S. Department of Energy (DOE), 1989 (October). Order 5400.4. *Comprehensive Environmental Response, Compensation, and Liability Act Requirements*.

DOE, 1991 (June). DOE/EH-0194. *Integrated Risk Information System*. Washington, DC: Office of Environmental Guidance.

DOE, 1992a (June). EH-231/012/0692. *CERCLA Baseline Risk Assessment*. Office of Environmental Guidance.

DOE, 1992b (December). EH-231/014/1292. *Use of Institutional Controls in a CERCLA Baseline Risk Assessment*. Office of Environmental Guidance.

DOE, 1993 (January). Order 5400.2A. *Environmental Compliance Issue Coordination, Change 1*.

Bowers, T.S., et al., 1996. "Statistical Approach to Meeting Soil Cleanup Goals." *Environmental Science and Technology*, Vol. 30, No. 5.

Federal Focus, Inc. 1991. *Toward Common Measures - Rrecommendations for Presidential Executive Order on Environmental Risk Assessment and Risk Management Policy*. The Institute for Regulatory Policy.

Gilbert, R. O. 1987. *Statistical Methods for Environmental Pollution Monitoring*. Van Nostrand Reinhold Publisher.

National Research Council (NRC), 1983. *Risk Assessment in the Government: Managing the Process*. National Academy Press/National Academy of Science. Washington, D.C.

NRC, 1994. *Issues in Risk Assessment*. Washington, D.C. National Academy Press/National Academy of Science.

A.2 RELATED PUBLICATIONS.

The following list represents additional sources of information related to this manual.

U.S. Army Environmental Hygiene Agency (USAEHA), 1992 (January). TG No. 185. *Commander's Guide to the Health Risk Assessment Process*.

U.S. Environmental Protection Agency (USEPA), 1983 (November). EPA/600/8-83/030. *Rapid Assessment of Potential Ground-Water Contamination Under Emergency Response Conditions*. Office of Health and Environmental Assessment.

USEPA, 1985 (August). EPA/560/5-85/026. *Verification of PCB Spill Cleanup by Sampling and Analysis*. Office of Toxic Substances.

USEPA, 1986 (June). OSWER Directive 9355.0-4A. *Superfund Remedial Design and Remedial Action Guidance*. Office of Emergency and Remedial Response.

USEPA, 1986 (May). EPA/560/5-86. *Field Manual for Grid Sampling of PCB Spill Sites to Verify Cleanup*. Office of Toxic Substances.

USEPA, 1986 (24 September). *Guidelines for Carcinogen Risk Assessment*. Office of Health and Environmental Assessment. 51 FR 33992.

USEPA, 1986 (24 September). *Guidelines for Health Risk Assessment of Chemical Mixtures.* Office of Health and Environmental Assessment. 51 FR 34014.

USEPA, 1986 (November). *Test Methods for Evaluating Solid Waste.* Office of Solid Waste and Emergency Response. SW-846. Third Edition.

USEPA, 1986. *Hazardous Waste Management Rule: Final Organic Leachate Model (OLM).* Office of Solid Waste and Emergency Response. 51 FR 41100.

USEPA, 1986. EPA/600/6-86/002. NTIS#: PB86-232774. *Development of Advisory Levels for Polychlorinated Biphenyls (PCB) Cleanup.* Office of Health and Environmental Assessment.

USEPA, 1987 (March). OSWER Directive 9285.4-02. *Guidance for Coordinating ATSDR Health Assessment Activities With the Superfund Remedial Process.* Office of Emergency and Remedial Response.

USEPA, 1987 (March). OSWER Directive 9355.7B. *Data Quality Objectives for Remedial Response Activities: Development Process.* Office of Emergency and Remedial Response and Office of Waste Programs Enforcement.

USEPA, 1987 (August). *Superfund Selection of Remedy.* Office of Solid Waste and Emergency Response.

USEPA, 1987. *Polychlorinated Biphenyls Spill Cleanup Policy.* Office of Pesticides and Toxic Substances. 52 FR 10688.

USEPA, 1988. OSWER Directive 9360.1-10. *Interim Final Guidance on Removal Action Levels at Drinking Water Contamination Sites.* Office of Solid Waste and Emergency Response; a memorandum from H.L. Longest to waste management division directors, Regions I-X and environmental services directors, Regions I, VI and VII.

USEPA, 1988 (May). OSWER Directive 9320.2-3a. *Procedures for Deleting Sites From the National Priorities List (NPL).* Draft. Hazardous Site Control Division, Office of Emergency and Remedial Response.

USEPA, 1988 (December). *Guidance on Remedial Actions for Contaminated Groundwater at Superfund Sites.* Office of Emergency and Remedial Response. EPA/540/G-88/003.

USEPA, 1989 (March 2). EPA/530-SW-89. *Guidance on Metals and Hydrogen Chloride Controls for Hazardous Waste Incinerators.* Draft. Prepared by Versar Inc. for the Office of Solid Waste, Waste Treatment Branch.

USEPA, 1989 (March). *Soil Sampling Quality Assurance User's Guide.* Environmental Monitoring Systems Laboratory.

USEPA, 1989 (April). SOW No. 2/88. *Contract Laboratory Program Statement of Work for Organics Analyses -- Multi-Media, Multi-Concentration.* Office of Emergency and Remedial Response.

USEPA, 1989 (October). EPA/540/2-89/057. *Determining Soil Response Action Levels Based on Potential Contaminant Migration to Groundwater: A Compendium of Examples.* Office of Emergency and Remedial Response.

USEPA, 1989 (October 21). *Managing the Corrective Action Program for Environmental Results: the RCRA Facility Stabilization Effort;* Memorandum from S.K. Lowrance and B.M. Diamond to Regions I-X Waste Management Division Directors. Office of Solid Waste and Office of Waste Programs Enforcement.

USEPA, 1989 (December 12). EPA/530-SW-90-021. *Report on Minimum Criteria to Assure Data Quality.* Office of Solid Waste.

USEPA, 1989 (December 13). *Guidance on Handling and Reporting Chemical Concentration on Data in Superfund Risk Assessment.* Smith RL et al. Technical Support Section, EPA Region III.

USEPA, 1989. OSWER Directive #9355.4-02. *Interim Guidance on Establishing Soil Lead Cleanup Levels at Superfund Sites.* Office of Emergency and Remedial Response.

USEPA, 1990 (January). OSWER Directive 9992.4. *Federal Facilities Hazardous Waste Compliance Manual.* Office of Waste Programs Enforcement.

USEPA, 1990 (April). OSWER Directive 9360.4-01. *Quality Assurance/Quality Control Guidance for Removal Activities: Sampling QA/QC Plan and Data Validation Procedures.* Interim Final. Office of Emergency and Remedial Response.

USEPA, 1990 (May). EPA/600/4-90/013. *A Rationale for the Assessment of Errors in the Sampling of Soils.* Environmental Monitoring Systems Laboratory.

USEPA, 1990 (July 18). *Hazardous Waste Management Rule: EPA CML Landfill Attenuation Factors (DAFs).* Final Rule. Office of Solid Waste and Emergency Response. 55 FR 32999.

USEPA, 1990 (July). *Health Effects Assessment Summary Tables and User's Guide.* Environmental Criteria and Assessment Office. Prepared for Office of Emergency and Remedial Response.

USEPA, 1990 (August). OSWER Directive 9355.4-1. *Guidance on Remedial Actions for Superfund Sites with PCB Contamination.* Office of Emergency and Remedial Response.

USEPA, 1990. EPA/530-SW-90-036. *RCRA Orientation Manual. 1990 Edition.* Office of Solid Waste/Permits and State Programs Division.

USEPA, 1991 (February). EPA/540/P-91/001. *Conducting Remedial Investigation/Feasibility Studies for CERCLA Municipal Landfill Sites.* Office of Emergency and Remedial Response.

USEPA, 1991 (24 April). Facsimile from John Wilson to Peter Tong. Subject: Biodegradation Rate or Extent of Removal of Chemicals in Pristine and Contaminated Subsurface Materials. U.S. Environmental Protection Agency, Office of Research and Development, Robert S. Kerr Environmental Research Laboratory.

USEPA, 1991 (May 26). *Implementing the Deputy Administrator's Risk Characterization Memorandum.* Memorandum from Henry Longest and Bruce Diamond to regional waste management division directors. Office of Emergency and Remedial Response and Office of Waste Programs Enforcement.

USEPA, 1991 (July). EPA/625/6-90/016b. *Handbook - Ground Water, Vol. II: Methodology.* Office of Research and Development.

USEPA, 1991 (August). EPA/625/6-91/026. *Stabilization Technologies for RCRA Corrective Actions.* Office of Research and Development.

USEPA, 1991 (November). *Implementation Document for Boiler and Industrial Furnace (BIF) Regulations.* Draft. Office of Solid Waste.

USEPA, 1991 (November). OSWER Directive 9360.4. *Removal Program Representative Sampling Guidance, Volume 1 - Soil.* Office of Emergency and Remedial Response.

USEPA, 1992 (January). EPA-450/1-92-002. *Air/Superfund National Technical Guidance Study Series: Guideline for Predictive Baseline Emissions Estimation Procedures for Superfund Sites.* Office of Air Quality Planning and Standards.

USEPA, 1992 (February 11). *New Interim Region IV Guidance.* EPA Region IV.

USEPA, 1992 (February). EPA-450/1-92-003. *Air/Superfund National Technical Guidance Study Series: Screening Procedures for Estimating the Air Impacts of Incineration at Superfund Sites.* Office of Air Quality Planning and Standards.

USEPA, 1992 (August). Intermittent Bulletin, Vol. 1, No. 1. *SACM Program Management Update - Identifying SACM Program Management Issues.* Draft. Superfund Revitalization Activity, Office of Emergency and Remedial Response.

USEPA, 1992 (August). *Assessing Sites Under Superfund Accelerated Cleanup Model - Quick Reference Fact Sheet.* Draft. Hazardous Site Evaluation Division, Office of Emergency and Remedial Response.

USEPA, 1992 (September). EPA-451/R-92-002. *Air/Superfund National Technical Guidance Study Series:*

Assessing Potential Indoor Air Impacts for Superfund Sites. Office of Air Quality Planning and Standards.

USEPA, 1992 (October). EPA/625/R-92/014. *RCRA Corrective Action Stabilization Technologies - Proceedings*. Office of Research and Development.

USEPA, 1992 (December 22). *Water Quality Standards; Establishment of Numeric Criteria for Priority Toxic Pollutants; States' Compliance, Final Rule*. Office of Water. 57 FR 60848.

USEPA, 1993 (September). *Data Quality Objectives Process for Superfund*. Interim Final Guidance. Office of Emergency and Remedial Response.

USEPA, 1993 (April). *Drinking Water Regulations and Health Advisories*. Office of Water.

USEPA. On-Line Database: *Integrated Risk Information System (IRIS)*.

American Conference of Governmental Industrial Hygienists, Inc (ACGIH), 1986. *Threshold Limit Values (TLVs) for Chemical Substances and Physical Agents and Biological Exposure Indices (BEIs)*.

Agency for Toxic Substances and Disease Registry (ATSDR), 1989. TP-88/10. *Toxicological Profile for Chromium*.

ATSDR, 1990. TP-88/17. *Toxicological Profile for Lead*.

Bailiff, M.D., and K.E. Kelly, 1990 (April). *Hexavalent Chromium in Hazardous Waste Incinerator Facilities: From Stack Emissions to Health Risks*. Paper presented at the American Waste Management Association International Specialty Conference on Waste Combustion in Boilers and Industrial Furnaces, Kansas City, MO; April 17-20, 1990.

Bennett, D., and P. Tong, 1988 (September). *The Use of Health/risk Assessment Information in CERCLA Related Activities*. Toxics Integration Branch, Office of Emergency and Remedial Response, U.S. Environmental Protection Agency.

Bradley, M.A., 1992. "RCRA Section 3019 Exposure Information and Health Assessments." *Environmental Permitting*. Winter 1992/1993. pp. 55-62.

Calabrese, et al., 1989. "How Much Soil do Young Children Ingest: Epidemiological Study." *Regulatory Toxicology and Pharmacology*. Vol. 10, pp. 123-137.

Chaney, R.L., 1985. PNSP/85-01. "Potential Effects of Sludge-borne Heavy Metals and Toxic Organics on Soils, Plants, and Animals, and Related Regulatory Guidelines." Annex 3, Workshop Paper 9, pp. 1-56. In: *Final Report of the Workshop on the International Transportation, Utilization or Disposal of Sewage Sludge Including Recommendations*. Pan American Health Organization.

Chaney, R.L., Sterrett, S.B., and Mielke, H.W. n.d. "Average ppm (dry) Heavy Metals in Soils from 422 Baltimore, Maryland, gardens." In: *The Potential for Heavy Metal Exposure from Urban Garden Soils*. Agricultural Research Service Biological Waste Management and Organic Resources Laboratory, U.S. Department of Agriculture.

Chemical Manufacturers Association (CMA), 1991a. *Analysis of the Impact of Exposure Assumptions on Risk Assessment of Chemicals in the Environment. Phase III: Evaluation and Recommendation of Alternative Approaches*. Exposure Assessment Task Group.

CMA, 1991b. *Analysis of the Impact of Exposure Assumptions on Risk Assessment of Chemicals in the Environment. Phase II: Uncertainty Analyses of Existing Exposure Assessment Methods*. Exposure Assessment Task Group.

DataMap®, Inc. Carrier Route Demographic Summary Reports - computer printout: CAM-1-Plus. Eden Prairie, MN.

Davis, S., and Waller, P., 1990. "Quantitative Estimate of Soil Ingestion in Normal Children Between the Ages of 2 and 7 Years: Population-Based Estimates using Aluminum, Silicon, and Titanium as Soil Tracer Elements." *Archive of Environmental Health*. Vol. 45, pp. 112-122.

Dragun, J., 1988. *The Soil Chemistry of Hazardous Materials.* Hazardous Materials Control Research Institute.

Hauchman F., 1991 (September). *Cancer Risks from Short Term Exposures Evaluated.* NATICH Newsletter. Pollution Assessment Branch, U.S. Environmental Protection Agency and STAPPA/ALAPCO.

Hawkins, N. C., 1991. "Conservatism in Maximally Exposed Individual (MEI) Predictive Exposure Assessments: A First Cut Analysis." *Regulatory Toxicology and Pharmacology* 14: 107-117.

Howard, P.H., et al., 1990. *Handbook of Environmental Fate and Exposure Data for Organic Chemicals. Volumes I and II.* Lewis Publishers, Inc. Chelsea, Michigan.

LaGoy, P., 1987. "Estimated Soil Ingestion Rates for use in Risk Assessment." *Risk Analysis.* Vol. 7, No. 3, pp. 355-359.

Lindsay, W.L., 1979. *Chemical Equilibria in Soils.* John Wiley and Sons Publisher.

MacDonnell, M.M., et.al., 1991 (September). ANL/CP--72945. *Strategy for Integrated CERCLA/NEPA Risk Assessments.* Argonne National Laboratory.

National Climatic Center (NCC), 1968 (June). *Climatic Atlas of the United States.* Environmental Science Services Administration, U.S. Department of Commerce.

Neptune, D., et.al., 1990. "Quantitative Decisionmaking in Superfund: A Data Quality Objectives Case Study. *Hazardous Materials Control.* May/June 1990 Issue.

Neptune, D., A.L. Morehead, and D.I. Michael, n.d. *Streamlining Superfund Soil Studies: Using the Data Quality Objectives Process for Scoping.* Quality Assurance Management Staff, Office of Research and Development; Environmental Research Planning Department, Research Triangle Institute.

National Research Council, (NRC), 1989a. *Recommended Daily Allowances.* Subcommittee on the tenth edition of the RDAs.

National Institute of Occupational Safety and Health (NIOSH), 1987 (August 15). *NIOSH Manual of Analytical Methods.* U.S. Dept. Health and Human Services. Third Edition, Revision #1.

NIOSH, 1990 (June). *NIOSH Pocket Guide to Chemical Hazards.* U.S. Dept. Health and Human Services.

Ruffner, J.A., 1985. *Climates of the States - National Oceanic and Atmospheric Administration Narrative Summaries, Tables, and Maps for Each State with Overview of State Climatologist Program. Volume 1.* Gales Research Company. Detroit, Michigan.

SRI International, 1982 (December). PB87-169090 *Aquatic Fate Process Data for Organic Priority Pollutants.* Prepared for U.S. Environmental Protection Agency, Office of Water Regulations and Standards.

Taylor. D., 1993 (January). Recycling sewage sludge: what are the risks? Health & Environmental Digest, Vol. 6, No. 9.

Verschueren, K., 1983. *Handbook of Environmental Data on Organic Chemicals. Second Edition.* Van Nostrand Reinhold Company. New York, New York.

Shacklette, H.T., and J.G. Boerngen, 1984. *Element Concentrations in Soils and Other Surficial Materials of the Conterminous United States.* USGS Professional Paper, U.S. Geological Service.

APPENDIX B
LIST OF ACRONYMS

ACSIM	Assistant Chief of Staff for Installation Management
ACL	Alternate Concentration Limit
AOC	Area of Concern/Area of Contamination
AR	Army Regulation
ARARs	Applicable or Relevant and Appropriate Requirements
ASA(I,L,E)	Assistant Secretary of the Army for Installations, Logistics, and the Environment
ATSDR	Agency for Toxic Substances and Disease Registry
BES	Biomedical Engineering Service
BRA	Baseline Risk Assessment
BRAC	Base Realignment and Closure
CAA	Clean Air Act
CDD	Chlorinated dibenzo-p-dioxin
CDF	Chlorinated dibenzofuran
CERCLA	Comprehensive Environmental Response, Compensation, and Liability Act
CERCLIS	CERCLA Information System
CERFA	Community Environmental Response Facilitation Act
CMI	Corrective Measures Implementation
CMS	Corrective Measures Study
COC	Chemical of Concern
COPC	Chemical of Potential Concern
CRC	Coastal Resource Coordinator
CRCB	Coastal Resource Coordination Branch
CSM	Conceptual Site Model
CT	Central Tendency
CWA	Clean Water Act
CX	Center of Expertise
DA	Department of the Army
DEP	Director of Environmental Programs
DERA	Defense Environmental Restoration Account
DERP	Defense Environmental Restoration Program
DL	Detection Limit
DOD	Department of Defense
DOE	U.S. Department of Energy
DON	Department of the Navy
DQO	Data Quality Objective

DSMOA/CA	Department of Defense and State Memorandum of Agreement/ Cooperative Agreement Program
ECAS	Environmental Compliance Assessment System
EE/CA	Engineering Evaluation and Cost Analysis
EM	Engineer Manual
EO	Executive Order
ERA	Ecological Risk Assessment
ESA	Endangered Species Act
EU	Exposure Unit
FFA	Federal Facility Agreement
FOSL	Finding of Suitability to Lease
FOST	Finding of Suitability to Transfer
FS	Feasibility Study
FUDS	Formerly Used Defense Sites
FY	Fiscal Year
HEA	Health and Environmental Assessment
HEAST	Health Effects Assessment Summary Tables
HHRA	Human Health Risk Assessment
HI	Hazard Index
HQ	Hazard Quotient
HQUSACE	Headquarters, U.S. Army Corps of Engineers
HRS	Hazard Ranking System
HSWA	Hazardous and Solid Waste Amendments of 1984
HTRW	Hazardous, Toxic, and Radioactive Waste
IAG	Interagency Agreement
IDL	Instrument Detection Limit
IEUBK	Integraged Exposure Uptake and Biokinetic Model
IRA	Interim Remedial Action
IRIS	Integrated Risk Information System
IRP	Installation Restoration Program
MC	Monte Carlo
MCL	Maximum Contaminant Level
MCLG	Maximum Contaminant Level Goal
MDL	Method Detection Limit
NAPL	Non-Aqueous Phase Liquids
NAAQS	National Ambient Air Quality Standards

NAS	National Academy of Sciences	SDWA	Safe Drinking Water Act
NCP	National Oil and Hazardous Substances Pollution Contingency Plan	SF	Slope Factor
		SI	Site Inspection
NFA	No Further Action	SITE	Superfund Innovative Technology Evaluation
NOAA	National Oceanic and Atmospheric Administration		
		SOW	Statement/Scope of Work
NON	Notice of Noncompliance	SQL	Sample Quantitation Limit
NPL	National Priorities List	SWMU	Solid Waste Management Unit
NRC	National Research Council	SVOC	Semi-Volatile Organic Compound
NTIS	National Technical Information Service	TARA	Technical Approach for Risk Assessment
OE	Ordnance and Explosives	TBC	To-Be-Considered
OMB	Office of Management and Budget	TCDD	Tetrachlorodibenzo-p-dioxin
OSWER	Office of Solid Waste and Emergency Response (USEPA)	TCL	Target Cleanup Levels
		TEF	Toxicity Equivalence Factor
OU	Operable Unit	TPH	Total Petroleum Hydrocarbons
		TPP	Technical Project Planning
PA	Preliminary Assessment	TSD	Treatment, Storage, or Disposal
PAH	Polycyclic Aromatic Hydrocarbon		
PbB	Blood Lead	UCL	Upper Confidence Limit
PCB	Polychlorinated biphenyl	UF	Uncertainty Factor
PM	Project Manager	USACE	U.S. Army Corps of Engineers
POL	Petroleum, Oil, and Lubricants	USACHPPM	U.S. Army Center for Health Promotion and Preventive Medicine
PQL	Practical Quantitation Limit		
PRG	Preliminary Remediation Goal	USAF	U.S. Air Force
		USEPA	U.S. Environmental Protection Agency
QA/QC	Quality Assurance/Quality Control		
QL	Quantitation Limit	UST	Underground Storage Tank
RA	Remedial Action	VOC	Volatile Organic Compound
RAGS I	Risk Assessment Guidance for Superfund		
RAO	Remedial Action Objective		
RBC	Risk-Based Concentration		
RCRA	Resource Conservation and Recovery Act		
RD	Remedial Design		
RFA	RCRA Facility Assessment		
RfC	Reference Concentration		
RfD	Reference Dose		
RFI	RCRA Facility Investigation		
RG	Remediation Goal		
RI	Remedial Investigation		
RMDM	Risk Management Decision-Making		
RME	Reasonable Maximum Exposure		
ROD	Record of Decision		
RPM	Remedial Project Manager		
SACM	Superfund Accelerated Cleanup Model		
SAP	Sampling and Analysis Plan		
SARA	Superfund Amendments and Reauthorization Act of 1986		

www.ingramcontent.com/pod-product-compliance
Lightning Source LLC
Chambersburg PA
CBHW051338200326

41519CB00026B/7472